No Longer the Property of
Hayner Public Library District

	DATE DUE	

Y970.004
SPA

Spangenburg, Ray

The American Indian
experience

HAYNER PUBLIC LIBRARY DISTRICT
ALTON, ILLINOIS

OVERDUES .10 PER DAY. MAXIMUM FINE
COST OF BOOKS. LOST OR DAMAGED BOOKS
ADDITIONAL $5.00 SERVICE CHARGE.

The American Indian Experience

The American Indian Experience

Ray Spangenburg and Diane K. Moser

HAYNER PUBLIC LIBRARY DISTRICT

American Historic Places: *The American Indian Experience*

Copyright © 1997 by Ray Spangenburg and Diane K. Moser

All rights reserved. No part of this book may be reproduced or utilized in any form or by any means, electronic or mechanical, including photocopying, recording, or by any information storage or retrieval systems, without permission in writing from the publisher. For information contact:

Facts On File, Inc.
11 Penn Plaza
New York NY 10001

Library of Congress Cataloging-in-Publication Data

Spangenburg, Ray, 1939–
   The American Indian experience / Ray Spangenburg and Diane K. Moser.
       p.   cm.—(American historic places)
   Includes bibliographical references and index.
   Summary: Describes historic places in the United States associated with Native American history and culture, including Little Big Horn Battlefield, Alcatraz Island, and Nez Perce National Historic Park.
   ISBN 0-8160-3403-6
   1. Indians of North America—History—Juvenile literature.
   2. Indians of North America—Antiquities—Juvenile literature.
3. Historic sites—United States—Juvenile literature. 4. National monuments—United States—Juvenile literature. [1. Indians of North America—History. 2. Indians of North America—Antiquities. 3. Historic sites. 4. National monuments.] I. Moser, Diane, 1944–
       II. Title. III. Series.
   E77.4.S63 1997
973—dc21   97-8387

Facts On File books are available at special discounts when purchased in bulk quantities for businesses, associations, institutions or sales promotions. Please call our Special Sales Department in New York at 212/967-8800 or 800/322-8755.

You can find Facts On File on the World Wide Web at http://www.factsonfile.com

Text design by Cathy Rincon
Cover design by Dorothy Wachtenheim
Layout by Robert Yaffe

Illustrations on pages vi, 5, 20, 33, 46, 49, 61, 66, 69, 77, 91 and 99 by Jeremy Eagle

This book is printed on acid-free paper.

Printed in the United States of America

RRD FOF 10 9 8 7 6 5 4 3 2 1

To the memory of

Claire Hampton Cox

1911–1996

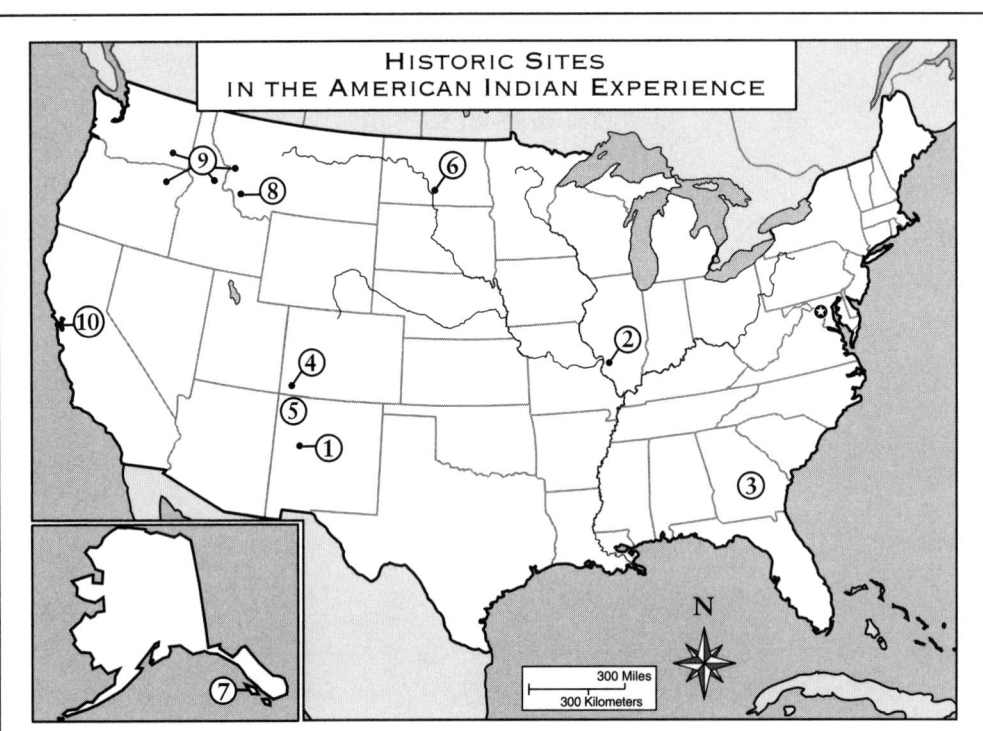

CONTENTS

Preface to the Series   IX

Acknowledgments   XI

Introduction   XIII

PETROGLYPH NATIONAL MONUMENT   1
Early Art of the Ancient Ones
Albuquerque, New Mexico

CAHOKIA MOUNDS STATE HISTORIC SITE   15
Remnants of an Ancient Metropolis
Collinsville, Illinois

OCMULGEE NATIONAL MONUMENT   25
Thousands of Years on the Banks of a River
Macon, Georgia

MESA VERDE NATIONAL PARK   39
Cliff Dwellings of the Mesa-Top Farmers
Mesa Verde National Park, Colorado

CHACO CULTURE NATIONAL HISTORICAL PARK 52
Great Pueblo City of the Southwest
Nageezi, New Mexico

KNIFE RIVER INDIAN VILLAGES NATIONAL 64
HISTORIC SITE
Earth Lodge Dwellers on the Plains
Stanton, North Dakota

SITKA NATIONAL HISTORICAL PARK 72
Where Tlingit Met Russian
Sitka, Alaska

LITTLE BIGHORN BATTLEFIELD NATIONAL MONUMENT 83
Fleeting Triumph over the United States Government
Crow Agency, Montana

NEZ PERCE NATIONAL HISTORICAL PARK 93
Homeland in the Northwest
Spalding, Idaho

ALCATRAZ ISLAND 107
Taking a Stand in the Twentieth Century
San Francisco Bay, California

More Places to Visit 116

More Reading Sources 121

Index 124

PREFACE TO THE SERIES

History doesn't have to be dry or stuffy. And it isn't exclusively about military skirmishes and legislative proclamations—they make up only a small part of it. History is the story of life events that happened to people who cared as passionately about their lives as we care about ours. And it's the story of events that often continue to shape and influence our lives today. But getting to the human side of these stories isn't always easy. That's why there's nothing like visiting the place where an event actually occurred to get the feel of what it all meant.

The study of historic places—what happened at a particular site and how the lives of the people there were affected—has emerged as a great way to approach history, to "relive" the experience, and open up to the immense diversity of American culture. Every community and region is rich in such places—places that highlight real stories about real people and events. Even if you can't actually visit such a place, the next best thing is to go there through pictures and words. Use this book and the other books in this series as jumping-off points and look around your community for places where you can experience the world of the people who once lived in your own region—and begin exploring!

◆ ◆ ◆

Each volume in this series explores a different aspect of U.S. history by focusing on a few select places. This book takes a look at historic places in the United States associated with Native American history and culture. Of course, choosing exactly which places to focus on in each book was one of the most difficult tasks of this project. We limited our choices to sites that had either been restored or maintained in authentic historic condition—most are National Historic Landmarks, chosen by the U.S. government to be preserved for their historic significance. We also tried to include examples from a variety of locations, events, and experiences, types of sites, and time periods. We then limited our selections to just a few. But many other fascinating places exist throughout the country, and that's why we mentioned other related sites at the end of some chapters (under Exploring Further) and added a list of additional sites at the back of the book (More Places to Visit).

Each chapter begins with information about the site (At a Glance). Then we explore the place—what it's like and who lived there, how the place relates to Native American history and culture, and what it's like to visit there today. We also look closely at one feature of the site in A Close-Up section, followed by a section recapping how the site came to be a protected historic site (Preserving It for the Future). A list of books and other resources concludes each chapter (Exploring Further), directing readers to either a broader or closer view of the American Indian experience.

Exploring historic sites not only provides a way to experience past events with fresh vividness and immediacy, it also offers a way of seeing the past through new eyes, through the eyes of those who lived it. For this adventure—and it can prove to be a lifetime adventure—this series will have accomplished its purpose if it provides the springboard for future explorations. In the words of an old Gaelic greeting, "May the wind be always at your back and may the road rise up to meet you" as you travel down these avenues of historical experience.

ACKNOWLEDGMENTS

Many people have given of their time, talents, resources, and enthusiasm to help us with this book, and we'd like to gratefully thank them all, including: Greg Gnesios of Petroglyph National Monument, Bill Iseminger at Cahokia Mounds, Jennifer Baird at Sitka National Historic Monument, Todd Eckert of the Golden Gate National Recreation Area, and Dennis and Anne Talbott. And at Facts On File, special thanks to Nicole Bowen, whose expertise and steady calm have seen us through, and to our former editor, James Warren, who helped us conceptualize the series.

INTRODUCTION

Long before the arrival of any European explorers, the land that we know today as North America was already settled by a large number of people, dispersed throughout its landmass. When Europeans such as the Vikings, Christopher Columbus, and Amerigo Vespucci first headed their ships across the Atlantic Ocean, they bumped unexpectedly into a continent they hadn't known was there. What's more, this "New World" turned out to be new only to them—because Native Americans of varied cultures and languages had been living there for many thousands of years.

Where had these indigenous people come from? The most likely explanation, for which there is a good deal of evidence, is that during the late Pleistocene geologic epoch (about 10,000 to 50,000 years ago), when humans migrated into Siberia, the formation of glacial ice coincidentally used up so much water that the Bering Sea had lowered. As a result, a bridge of land beneath the sea became exposed, allowing animals and people with all their belongings to walk from Russia to Alaska.

"This much is certain, the first Americans were *Homo sapiens* who came from northeast Asia via the Bering Straits," anthropologist David J. Meltzer has stated, "by 11,500 years ago, hunter-gatherers had arrived here in time to see the end of the Pleistocene Era. Beyond those bare facts," he continued, "there is controversy."

Most likely, the people just followed the caribou and other animals that they were tracking, and then continued following their game straight on into the North and South American interior.

All this happened many thousands of years B.C., and after the Pleistocene came to an end, the Bering Sea rose again, and those who had used the land bridge were in North America to stay. Most never wanted to leave—the hunting was good, the fishing was good, and in some areas even farming (which they learned how to do) was good.

Archaeologists have found evidence of human habitation in areas as far apart and diverse as Russell Cave (9000 B.C.) in Alabama and Petroglyph National Monument (which may date back to 10,000 B.C.) in New Mexico. In New Mexico, the first Americans began creating petroglyphs and pictographs—pictures carved or painted on stone—as early as 800 B.C. Along the Scioto River in southern Ohio, the remarkable Hopewell people built great earthen mounds covering as many as 100 acres, as did a number of others of the early Woodland period (about 500 B.C. to A.D. 400). The Hopewells disappeared, practically without a trace, by A.D. 550. But mound building by no means stopped—the mound builders of the great city of Cahokia on the Mississippi began building around A.D. 700 and so did mound builders in Ocmulgee, on the Macon Plateau, around 900.

In the Southwest, the people the Navajo call the Anasazi built great cities of sandstone from as early as—or earlier than—A.D. 600, with the building of pueblos (adobe, or mud-brick, towns) in Chaco Canyon, New Mexico and in the caves and canyon walls of Mesa Verde, Colorado and elsewhere throughout the Southwest.

These were just a few of the early civilizations that grew and flourished long before European Americans appeared on the scene. By the time Hernando De Soto came from Spain in 1539–42, many civilizations had already come and gone, many nations had grown up, and many languages and divergent cultures had developed. The people found by European explorers had vastly diverse histories, customs, and religions—a rich heritage stretching back thousands of years.

Native Americans had already built the first multistoried apartment buildings in the Americas, carved the largest Native wood carvings known, dug 16-mile-long irrigation systems for farming in desert areas, constructed enormous earthen mounds up to 10 stories high, and developed democratic social systems.

But Europeans misperceived this, for the most part, viewing the American "Indians," as they mistakenly named them, in one of two ways, both incorrect. Some Europeans saw Native Americans as wild, uneducated, unsophisticated, and ferocious savages. Others thought of them as idealized children of nature, the "Noble Savages" described by Jean-Jacques Rousseau, Daniel Defoe, and others.

As the United States struggled to form a new nation, the country's repeated government failures to treat Native peoples with the respect due independent nations resulted in grievous confrontations. Moreover, by the end of the 19th century, the tribes had been overrun for the sake of gold rights, farmland, and railroad rights of way. They had been tricked or coaxed or placed by force onto reservations as small as a tenth the size of the lands they once had roamed.

The Indian began to see the white intruders as godless, dishonest, greedy, and not to be trusted. For those who did trust, and were generous to the whites, the deadly scourge of European diseases often wiped out hundreds of their people.

Clichéd thinking has often kept mainstream American culture from appreciating the diversity present among American Indian cultures. The stereotype of the Plains Indian warrior, with flowing feather headdress, beaded clothing, fiery pony, and coup stick or tomahawk may evoke a love of freedom and even violence that many Americans have grown up admiring and envying. But it is a limited and unreal stereotype. American Indians are as varied in their cultures as the terrain they inhabit, from the fishers along the Northwest Coast, to the farmers in the Southeast and in the Southwest, the buffalo hunters of the Plains, and the subsistence hunters of the northern Subarctic. They have lived in many dramatically different climates—from desert to prairie grassland to the oak forests of the Atlantic coast and the tropical swamplands of southern Florida. Divided into more than 300 different tribes, Indians have developed many vastly different cultures, and hundreds of different languages.

By listening carefully, one can learn much from Native Americans about harmony, an integrated universe, and a concern for resources necessary to existence, as well as an understanding of human responsibilities that are only now beginning to be recognized by the industrialists and politicians of the technological societies.

But above all, the Indians of North America always have been and still are survivors. Each tribe, throughout its long history, has adapted to and assimilated many other cultures—always conscious of its own sense of history and culture and place in the universe. It is a characteristic central to Native American thinking, as the descendants of those who crossed the Bering Sea long ago seek to carry forward into the modern world the legacy of their ancient beliefs and ideals.

Petroglyph National Monument

EARLY ART OF THE ANCIENT ONES
Albuquerque, New Mexico

AT A GLANCE

Evidence of human habitation from: ca. 10,000 B.C.

Petroglyphs dating from: ca. 800 B.C. to A.D. 1500

Ancient drawings etched in rocks on the mesas and along the canyon cliffs

Over 7,100 acres in size, the Petroglyph National Monument occupies a region west of Albuquerque extending from Piedras Marcadas Canyon in the north to Mesa Prieta in the south, and from five volcanoes in the west (Butte Volcano, Bond Volcano, Vulcan Volcano, Black Volcano, and JA Volcano) to the escarpment edge in the east. It preserves some 15,000 ancient drawings on stone, known as petroglyphs, created by ancestors of the Pueblo people.

Address:	**Comanaged by:**
Petroglyph National Monument	City of Albuquerque Open Space Division
4375 Unser Boulevard, NW	P.O. Box 1293
Albuquerque, NM 87120	Albuquerque, NM 87103
(505) 839-4429	(505) 873-6620

> High on the windswept mesa and steep slopes of New Mexico, these timeless images carved in stone convey haunting echoes of a distant past.

Honor. Preserve. Live each day with a good heart,
a clean soul, so that everyone you come in touch with
feels the same way.
—A Pueblo elder

♦ ♦ ♦ ♦

Among the ruins of the great pueblos that once stood in the canyon now called Piedras Marcadas ("Marked Rocks") lie the remains of petroglyph murals—drawings of stars, clouds, and other figures—that once lined their walls. (Greg Gnesios. Courtesy of the National Park Service)

Sweeping geologic formations, the stark play of light and shadow, subtle variations of color, and beautiful life-forms create a natural splendor in this region that alone is hard to resist. But walking the trails of Petroglyph National Monument you encounter an additional beauty: symbols left behind by a people who lived long ago, close to the land, in a manner so far removed from our lives that we can only attempt to imagine it. Here, shaman images and clown figures, four-pointed stars, animal shapes, shields, and masks speak to us from far in the past in ways we may intuitively respond to but can never completely understand. For Pueblo descendants of those who drew these petroglyphs (drawings in stone, from the Greek words *petra*, "rock," and *glyph*, "carving"), they have special meaning—they are powerful cultural symbols reflecting the complex society and religion of the Pueblo peoples of long ago, a link with ancestors, with shared parables, with a way of life. For contemporary Indians, this monument represents a hallowed landscape, a place where sacred ceremony still takes place. For visitors who do not have this special link, the petroglyphs evoke a universal human heritage. While they telegraph no direct messages—they are not hieroglyphics or an ancient form of writing—they convey the evocative meaning of art and symbolism. For everyone, it is an area requiring respect and care.

More than 200,000 years ago, long before the first humans arrived in this region just west of what is now the city of Albuquerque, New Mexico, the earth's crust bulged and broke, forming a series of rifts, or cracks, where molten lava broke through. This rift zone, now concealed by layers of sediment, extends along 400 miles of the Rio Grande and, in the Albuquerque area, it is 30 miles wide and extends several miles deep. Waves of lava broke through here and, at first very fluid, streamed eastward, coating the earth beneath with thin layers. Then violent gushes of more viscous lava burst upward, forming five volcanic cones in the area. More recently, erosion has eaten away at the sediment beneath the lava, breaking off boulders of basalt rock and forming a sharp escarpment, or cliff. West of this escarpment lies what is known as the West Mesa. The surface of the basalt rocks of this region has a dark patina, or varnish, sometimes black, sometimes brown, sometimes purple. The coating is caused, scientists think, by minerals left behind by water seeping over the rocks.

Some 12,000 years ago, when the first people came to the West Mesa of what is now known as the Petroglyph National Monument, the weather was very different from the dry, desert plateau we see today. Plentiful rains formed shallow lakes, and the green of lush plant life sprang up all around. Small animals, birds, mastodons, mammoths, ancestors of the bison, and even camels roamed the region in abundance. Here the earliest ancestors of the Pueblo Indians hunted and gathered food near the lakes to the west of the volcanoes.

From the stone of the western edge of the mesa above Rio Puerco, they chiseled spear points and stone scrapers to use on hides. They built camp fires and as the seasons changed, they followed the animals they hunted. But then the climate began to change. The weather became drier, the grasslands less lush. Slowly the game these people hunted roamed farther and farther away or died off, and these early ancestors began to develop a new way of living, a less nomadic "Archaic" culture.

Now, instead of following the big game, the hunter-gatherers went after smaller creatures, such as rabbits and deer, and they picked berries and dug up nutritious tubers. Archaeologists have found evidence on West Mesa that these hunter-gatherers wore down large stone surfaces called metates by grinding seeds with stone manos they held in their hands. They have found remains of cooking pits where they steamed vegetables and roasted meats, and post holes have marked the spots where they built places to live.

On the mesa above Rinconada Canyon, 5,300 feet—about a mile—above sea level, the nighttime temperature may dip to freezing in summer or winter. The land looks craggy and harsh, and yet, here the very earth incubates life. The basalt of the mesa absorbs the heat of day while holding water, producing a longer growing season than nearby desert areas, and plants and small animals flourish. The ancestors of the Pueblo people began to grow corn in the arroyos (or gullies) of the escarpment below Rinconada Mesa. They placed rocks in rows to slow the rush of water and soil down the hillsides and they built garden terraces, of which visitors can still see traces, along with chips of broken pottery and the outlines of small rooms in which they lived. These people foraged for food, but by sometime between 800 B.C. and A.D. 400 the people in the Rio Grande area had also become farmers, attached to the land. They settled in, building walls and digging pits to store harvested food. They developed rituals and prayers, hoping to encourage rain and

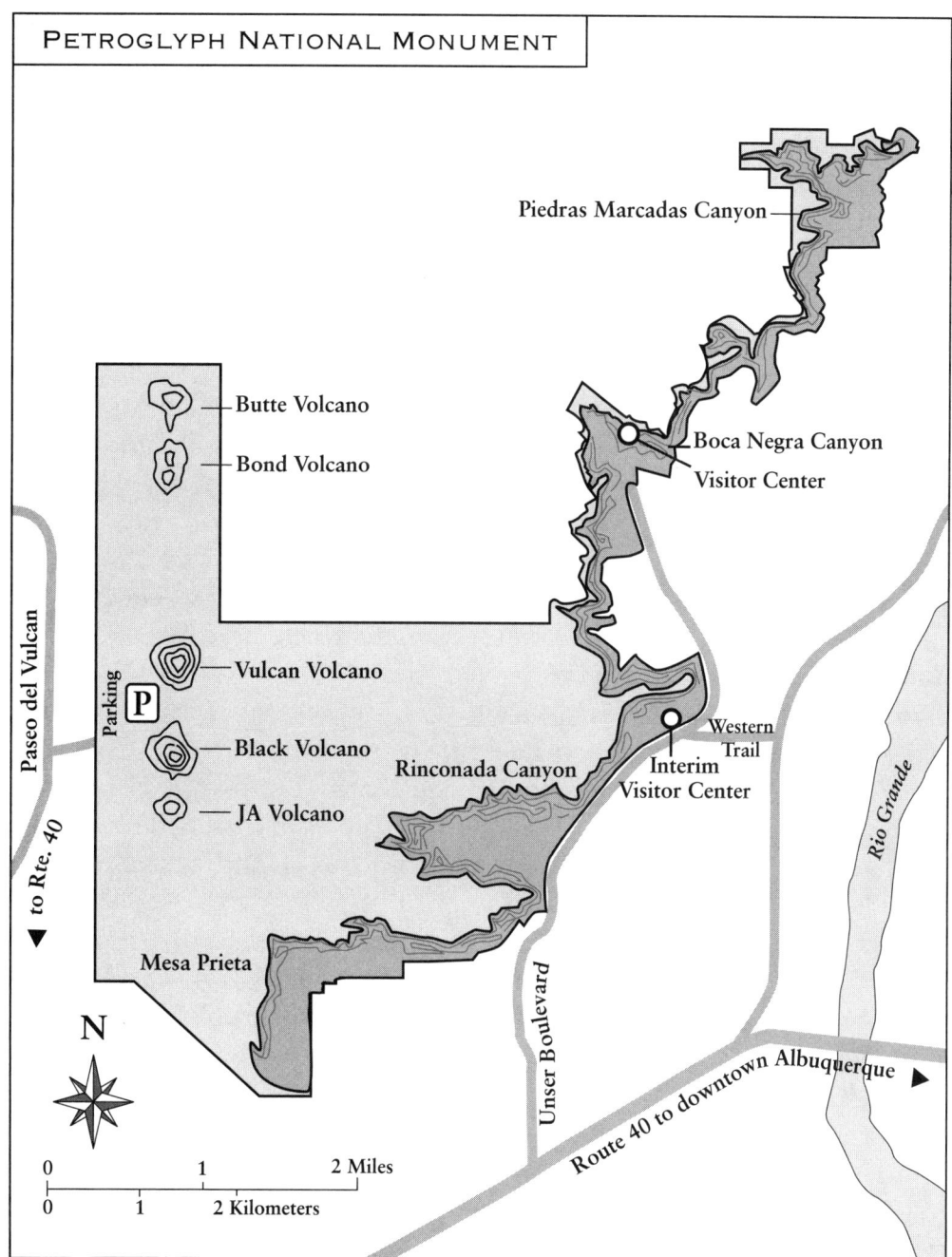

A short drive from Albuquerque, Petroglyph National Monument is the only national park devoted entirely to the preservation of ancient rock art created by the first inhabitants of the Southwest.

bountiful harvests. Archaeologists believe that during this time these people may have drawn some of the earliest of the petroglyphs found on the escarpment, or steep slope, of this area.

Working with a rock hammer and perhaps a stone used as a chisel, they scratched through the dark outer layer of basalt boulders, creating designs composed of circles, rakes, and wavy lines. What did these drawings mean? The circles may have represented the moon or sun, whose cycles were so key to the survival of these early farmers. Or, some interpreters have thought they might represent the complete cycle of a year or an opening to some other world. Because time in the Pueblo worldview is cyclical, not linear, these circles could also represent a beginning, since any point along the circumference of a circle might be a beginning.

No one is very sure about the meaning of the other forms of these earliest symbols—wavy lines may represent lightning or the flow of water, and the rakelike symbols may be methods of counting, falling rain, or rays of sun. They may represent these natural elements or something more, or, as is often the case with symbolism, they may stand for many ideas at once. Of one thing archaeologists seem sure, however: Chiseling these drawings took time, and they could not have been created haphazardly.

Dating the petroglyphs is difficult, and archaeologists use clues and correspondences among various sites where petroglyphs are found. They may match one style with another, found among charcoal or pottery or building structures that are more easily dated. Researchers have classified these early circles and wavy lines with a period style called Great Basin Abstract, found also in Nevada and eastern California.

A second, later group of petroglyphs seems to correspond to a period from about 700 to 900, known as the Pueblo I phase. Typical of this period is an outlined cross, found throughout the Southwest among Pueblo cultures, very likely representing the four directions: sunup and sundown (east and west) bisected by north and south. Depending upon the time of year, the directions shift. Division of circles and surfaces into four quadrants in this way still pervades Pueblo design.

During the following period, from about 900 to 1300, more images emerge in the petroglyphs—small animals, figures that look human, spirals, and prints of hands and feet. We can see that both hunting and farming remain important in the lives of the Pueblo people during this time. Game

Animals common to the landscape, such as this deer, often appear in petroglyph drawings. (Greg Gnesios. Courtesy of the National Park Service)

animals, tracks, and bows and arrows etched in the rocks show that these pursuits continue to be on the minds of the rock carvers. At the same time, the concerns of an agricultural people are evident: Will the seasons be favorable to growing crops? Will the sun shine and the rain fall as needed? Will the seeds grow? Will the harvest be sufficient to last through the long winter? Symbols of fertility abound, such as snakes and spirals, as well as depictions of rain, the sun, and triangular clouds.

Some petroglyphs may have had ceremonial significance, suggesting an effort to maintain balance between respect for the forces of nature and the demands of an emerging, ever more complex human society. Much is beyond human control, these symbols seem to say—yet they appear to reflect a belief, which remains part of Pueblo culture today, that if people respect nature's forces, and approach them responsibly, then they may succeed in living in harmony with the world around them.

From the time of the 14th century, the petroglyphs reflect the effects of a great cultural mixing in the valley of the Rio Grande. The Pueblo traded widely, and seashells from as far away as the Gulf of Mexico and the shores of California have been found here. In addition, an ever-worsening drought encouraged peoples from all over the Southwest to migrate to the fertile river valley around the West Mesa. Large pueblos lined the canyons of the West Mesa escarpment, and among them one now named Piedras Marcadas ("Marked Rocks"). The pueblos extended several stories high and consisted of a thousand rooms, surrounding two central plazas. Only ruins remain of the structures, but we can tell that the walls of the ceremonial rooms were covered with colored petroglyph murals portraying ceremonial figures, including mountain lions, four-pointed stars, and clouds. Archaeologists can verify that the style of these figures matches the style of images in murals, pottery, and stone drawings found in other areas of New Mexico as much as three centuries earlier. They believe that the appearance of this style here in the 1300s corroborates the idea that cultural mixing and the challenges of the drought had encouraged the development of shared beliefs and concepts across wide areas.

By this time the petroglyphs become more diverse and include shaman figures, clan symbols, and clown figures (representing an important aspect of Pueblo culture). Imposing shields and masks appear, decorated with patterns of feathers, lines, circles, and other striking designs. Supernatural characters surface, figures from

Later drawings, such as this one from Rinconada Canyon, take the form of sophisticated portrayals of shamans, mythological figures, and figures wearing masks and headdresses. (Greg Gnesios. Courtesy of the National Park Service)

Pueblo morality tales dance among the rocks, and kilted fluteplayers appear to recite their melodies. These figures are much more complex, intricate, and dynamic than the earlier ones.

In these petroglyphs the artist may convey an idea through the placement of a figure—for example, portraying watchfulness or omniscience with a head drawn on the corner of a jagged boulder, with one eye facing one way and the other looking out from around the corner. And archaeologists may ascribe a ceremonial context from the smooth, indented surface of a nearby stone where medicinal herbs may have been ground.

Birds abound in the petroglyphs, as they do in nature today throughout the region. Parrots interlock, eagles soar, and hawks glide through the sky. Brightly colored macaws are everywhere. In addition, there are lizards, snakes, and fish. And groups of different images appear—snakes, birds, turtles, cougars, frogs, fireflies—creatures the Pueblo saw and knew, that roamed the hills with them and stalked the night.

And the Pueblo added their imaginations and visions of these creatures, made emblems of them, and symbols. There are men shaped like turtles with rounded bodies and stubby arms and legs. Sometimes images are fused into strange combinations, such as a star-mask or a headdress with bird talons. Or they are exaggerated, such as a huge foot at the end of an elongated leg. Natural features of the stones are incorporated into the drawings to become noses or mouths.

By the 16th century, many of the Pueblo people had abandoned the pueblos by the river and moved on to greener lands. But thousands still remained when the Spanish conquistadores arrived in the valley of the Rio Grande and claimed the land for the king of Spain. In return, the king granted them huge tracts of land, which included the West Mesa and its escarpment. Many of the Pueblo people abandoned their homes, but the conquistadores made slaves of some, while Catholic missionaries sought to convert them. After four generations of oppression, in 1680 the Pueblo drove out the Spanish, who in turn reconquered New Mexico 12 years later. The land of the West Mesa became part of a land grant given to a soldier named Fernando Duran y Chaves, whose family had been among the earlier settlers.

Throughout the West and Southwest, the intermingling of Spanish Catholic and Native American cultures has left a legacy that has affected both traditions. The Pueblo in this region often took care of Spanish

Human hands are etched into the rock in this petroglyph. (Greg Gnesios. Courtesy of the National Park Service)

children and taught them much of their culture. They in turn assimilated some Christian beliefs and Catholic rituals. But, except for a few later drawings of Christian crosses left, perhaps, by Pueblo herdsmen, little of this era appears in the petroglyph record.

Instead, the petroglyphs speak of an earlier age, a time closer to the beginning of all time, evoked by the creation stories of the Pueblo that tell us people once had tails like lizards and all creatures could talk, a time when, perhaps, people felt more at one with their surroundings. As we walk as visitors among these drawings in the burnished rocks outside Albuquerque—the lights of a modern city not far away—we can feel the presence of this people of long ago and we can respect and honor their memory.

A CLOSE UP — ANCIENT ART WORKED ON STONE

The ancient drawings we see at Petroglyph National Monument and most other locations in the Southwest are called *petroglyphs* because they are created, not by adding paint or charcoal or dyes, but by removing a portion of the rock—through carving, scratching, "pecking," or chipping with a pointed piece of harder rock. Sometimes, the artist has scraped or "grooved" the rock face. The pecking tool could be held in the hand and wielded like an ice-pick, or held like a chisel in one hand while striking it with a "hammer-stone" held in the other hand. The hammer stone method provides much better control and, considering the intricacy of some of the drawings and the difficulty of carving into rock, it was probably the method most frequently used.

Another art form, known as pictographs, is the second major category of ancient rock art. Pictographs are more like paintings created on the rocks by using pigments, usually obtained from mineral sources. Sometimes, ancient artists combined the two techniques. But pictographs tend to fade with weathering, and they are usually well preserved only in well-sheltered locations, such as caves or shallow alcoves. If pictographs once were drawn on the rocks of Petroglyph National Monument, the long exposure of the rocks to intense weather conditions has long since caused them to vanish.

PRESERVING IT FOR THE FUTURE

Preservation of the petroglyphs in this area began at Boca Negra ("Black Mouth") Canyon, thanks to a 1960s campaign by a group of private citizens and geology students from the University of New Mexico. The result was the establishment of Indian Petroglyph State Park and Albuquerque's Volcano City Park. By 1986, the Las Imagenes National Archaeological District—which includes the entire escarpment and its petroglyphs—was added to the National Register of Historic Places.

In 1990, Congress established Petroglyph National Monument to preserve the more than 15,000 petroglyphs that have been found here. Additionally, the site preserves the natural resources of more than 7,100 acres of land, including biotic habitats in the mesa-top grassland, juniper savanna, the escarpment area of cliffs and tumbled boulders, mixed shrub grasslands, and desert willow wash.

The park is the only unit in the National Park System that is dedicated to the preservation of "rock art," as the petroglyphs are sometimes called. And its establishment specifically recognizes the importance the petroglyphs represent for the preservation of Native American cultural and spiritual values.

The monument is managed by a consortium of the City of Albuquerque Open Space Division, the State of New Mexico Park and Recreation Division, and the National Park Service, with an advisory board composed of representatives from business, the Native American community, the scientific community, and the general public.

Park curators and rangers caution that this historic site is unusually fragile—requiring the help and respect of visitors for its protection. Because petroglyphs are delicate and cannot be replaced, visitors are asked not to touch the images, make tracings, or otherwise disturb the rock surfaces.

Two visitors' centers have been set up—one north of Rinconada Canyon, at 4735 Unser Boulevard, NW, and the other at Boca Negra Canyon, at 6900 Unser Boulevard, NW, just north of Montaño Road. Three self-guided trails extend from the Boca Negra Canyon center.

EXPLORING ◆ FURTHER

Books about Petroglyphs and Pueblo Indians

Baylor, Byrd. *When Clay Sings*. Illustrated by Tom Bahti. New York: Scribner, 1972.

Lamb, Susan. *Petroglyph National Monument*. Tucson, Ariz.: Southwest Parks and Monuments Association, 1993.

Lister, Robert H., and Florence C. Lister. *Those Who Came Before*. Tucson, Ariz.: Southwest Parks and Monuments Association, 1983.

Muench, David (photographer). *Images in Stone*. Text by Polly Schaafsma. San Francisco: BrownTrout, 1995.

Ortiz, Alfonso, ed. *New Perspectives on the Pueblos*. Albuquerque, N.M.: University of New Mexico Press, 1972.

Petersen, David. *The Anasazi*. New True Books. Chicago: Children's Press, 1991.

Schaafsma, Polly. *Indian Rock Art of the Southwest*. Albuquerque, N.M.: University of New Mexico Press, 1980.

Spielmann, Katherine. *Farmers, Hunters, and Colonists.* Tucson, Ariz.: University of Arizona Press, 1991.

Tyler, Hamilton. *Pueblo Animals and Myths.* Norman, Okla.: University of Oklahoma Press, 1975.

Young, Jane. *Signs from the Ancestors.* Albuquerque, N.M.: University of New Mexico Press, 1988.

Related Places

Russell Cave National Monument
Route 1, Box 175, 3729 County Road 98
Bridgeport, AL 35740
(205) 495-2672
8 miles west of Bridgeport, Alabama (From U.S. Highway 72 at Bridgeport, follow Jackson County 75 West to Jackson County 98)

Artifacts found in this ancient cave indicate that Native Americans lived here for about 9,000 years, spanning three great eras: the Archaic period, from at least 7,000 B.C. to 500 B.C.; the Woodland period, from 500 B.C. to A.D. 900; and finally, the Mississippian period, from 1000 to about 1540. Depending on availability of park staff, upon request visitors may see a demonstration of archaic weapons and tools. Guided tours can be prearranged for organized groups.

Three Rivers Petroglyph Site
28 miles south of Carrizozo on U.S. Highway 54,
5 miles east on County Road B30
Three Rivers, NM 88352
(505) 525-4300 Bureau of Land Management, Las Cruces, NM

It's an easy 1,400-yard hike from this site's picnic area to some of the more interesting petroglyphs. Here, around 21,000 images have been recorded within a small, .22-square mile area. Park rangers caution visitors to remember that this part of southern New Mexico can be hot during the summer and fall, so bring plenty of water. Across the road from the picnic area is a small village site, consisting of one multiroom adobe structure, a masonry house, and a pithouse. The site has been partially excavated, giving a glimpse of life among the Jornada branch of the Mogollon culture.

Capitol Reef National Park
HC 70, Box 15
Torrey, UT 84775
(801) 425-3791
E-mail: care_superintendent@nps.gov

This remote area of Utah, in the Fremont River Canyon, has only become accessible since paved construction of Utah Highway 24. Earliest traces of human activity here—of a group known by archaeologists as the Fremont culture—date from the ninth century. They left behind many petroglyphs, probably created between 700 and 1275, a large number of which have weathered the elements along the cliff. Apparently related to the Four Corners Anasazi people, but less sophisticated, the Fremont Indians suddenly abandoned their settlements and fields in the 13th century. No explanation has ever been found.

Cahokia Mounds State Historic Site

REMNANTS OF AN ANCIENT METROPOLIS
Collinsville, Illinois

AT A GLANCE

Earliest settlements established: ca. 9500–8000 B.C.

City of Cahokia established and mounds built: ca. A.D. 700–1400

Remnants of a large prehistoric Native American community

This site preserves 68 of the 104 recorded mounds in this area that are the most striking remaining indications of a large center of commerce near the convergence of the Mississippi and Missouri rivers, across the Mississippi River from what is now St. Louis, Missouri.

Address:
Cahokia Mounds State Historic Site
P.O. Box 681
Collinsville, IL 72234
(618) 346-5160; fax, 346-5162

> From Cahokia on the Mississippi River, trade routes extended throughout most of North America, and in its day no other settlement north of Mexico equaled this great city's importance.

> Though I have worked at Cahokia for 25 years, I still marvel at what I see. It is an awesome site, massive and mysterious, especially in the predawn hours as I drive past the dark shapes of mounds poking through ground-hugging mist . . .
>
> —William R. Iseminger, archaeologist and curator,
> Cahokia Mounds State Historic Site

♦ ♦ ♦ ♦ ♦

The time of the annual harvest festival has arrived, and Cahokians are preparing for their annual celebration of the fall equinox, when day and night are equal in length and the long days of summer's growing season are past. From far distant places, traders have arrived with their goods to offer to the mighty lords of Cahokia. The goods are many and as varied as the places they come from—beads and ornaments of copper, drinking vessels made from conch shells, shellwork made into baskets, rare pigments from ocher and hematite for coloring pottery or for dying clothing or for use as body paint. The offerings seem endless. In exchange the Cahokians offer the wealth from their fields—squash, corn, pumpkin, seeds—and goods they have fashioned—woven fabrics, feathered capes, and garments of otter fur or mink or beaver. The great harvest market of Cahokia will send these and other goods back along the trade routes to far places, carrying the symbols of Cahokia's power—black ceramic vessels bearing carefully incised designs, including nested chevrons, forked eyes, and interlocking scrolls.

Recognized as the most sophisticated prehistoric native civilization north of Mexico, the city of Cahokia, as we have come to call it, is located in western Illinois, across the Mississippi River from what is now St. Louis. Home to

Painting showing what Cahokia may have looked like at the peak of its power. The 40-acre Grand Plaza, bounded by a wooden stockade, was central to a community of elite residences and temples, dominated by Monks Mound at one end and the Twin Mounds at the other. Other residences lay outside the stockade—perhaps housing less wealthy Cahokians, as well as some elite. Near the edge of this community is the circle of poles we call Woodhenge, where priests may have led rituals or ceremonies during observance of solstice and equinox sunrises. (William R. Iseminger. Courtesy of Cahokia Mounds State Historic Site)

approximately 20,000 residents at its height, Cahokia was once the greatest power in the center of the North American continent.

The broad, fertile floodplain of the Mississippi had been home to Indians for hundreds of years, living as hunter-gatherers off the plentiful plant and animal life they found all about them. Later, they began cultivating crops. Then, beginning around A.D. 700, the Indians of the Late Woodland culture inhabited the area for about 150 to 200 years. They lived in compact villages, hunted, fished, gathered wild foods, and planted seed-bearing crops such as corn.

Then another, more sophisticated culture emerged between 850 and 900. Probably descendants of the Late Woodland people, this group, known as the Mississippian culture, used more sophisticated technologies to improve their agricultural system, adding corn and squash to their cultivated crops.

The bountiful and stable food base of the region made possible the development of the city we call Cahokia, which became a regional center for a group of outlying communities, many of them as far away as Macon, Georgia (see Ocmulgee Mounds National Monument). Closer at hand, communities existed where St. Louis, Missouri; East St. Louis, Illinois; Lebanon; Mitchell; and Dupo are located today.

In part as a symbol of its power, the people of Cahokia erected, between the years 700 and 1400, 120 earthen mounds, moving 55 million cubic feet of earth in the process. Mound building took place all over the eastern, southeastern, and Mississippi Delta regions of what is now the United States, beginning in about A.D. 500. The mounds took many shapes—rectangular, or conical, or ridge top shapes. Or some builders constructed mounds in the shape of effigies—animal shapes, for the most part, such as the great serpent

Monks Mound and the Twin Mounds as they appear today (Courtesy of Cahokia Mounds State Historic Site)

Monks Mound—the largest mound at Cahokia—contains 22 million cubic feet of earth and is 100 feet high. At its base it covers 14 acres. (Courtesy of Cahokia Mounds State Historic Site)

mound that winds across the fields at Hopewell Culture National Historical Park near Chillicothe, Ohio.

But Cahokia's mounds were both numerous and large. One of them, Monks Mound, was as tall as a 10-story building, the largest prehistoric earthen structure in the Western Hemisphere. The entire project was completed using stone tools, the builders carrying some 15 million basketfuls of dirt on their backs, a task that would take 100 men 50 years to complete, working nonstop. On top of the mound, they probably also built structures (now gone) to be used either as a residence for the highest Cahokian officials, or possibly as a burial preparation site.

Today, fewer than 80 of the mounds at Cahokia remain (68 within the state park), but you can still see the giant Monks Mound, the plaza that stretched in front of it (now bisected by landscaping), and the Twin Mounds that used to reside across the great plaza in which Cahokians met and played a spectator sport similar to lacrosse, as well as a game that involved throwing

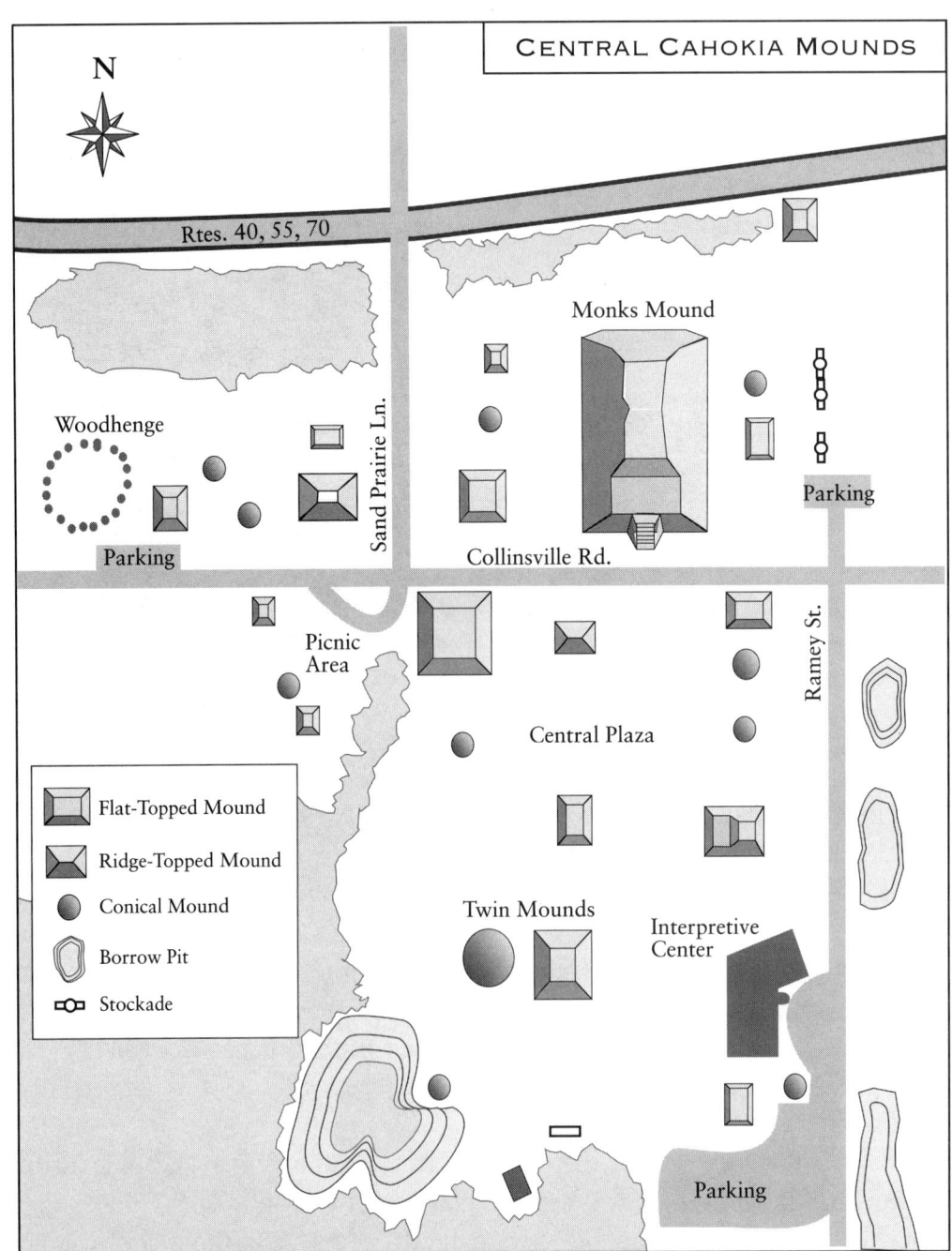

Map of Cahokia Mounds State Historic Site, showing Woodhenge on the eastern edge, Monks Mound, the Twin Mounds, the Grand Plaza, and other nearby sites.

THE AMERICAN INDIAN EXPERIENCE

a stone followed rapidly by a lance thrown at the spot the stone would most likely land.

The Cahokians fortified their city by a 2-mile long stockade fence, which probably indicates that there were threats from outside, although the stockade may have been a social barrier, with the elite housed within the fence and the less well-to-do citizens outside. The fence, built between 1100 and 1250, was a project in itself, requiring 15,000 to 20,000 logs, 1 foot in diameter and 20 feet tall.

The Cahokians had a colorful cultural life—probably including ceremonial and symbolic dancing accompanied by rattles, drums, and a flutelike instrument of which archaeologists have found remains. Both men and women wore body paint (the pigments for which came through their trade routes).

However, by the 1200s—about the time the Crusades were ending in Europe—Cahokia's power and influence began to decline, and by 1400 the great Cahokian culture appears to have just faded away. Like the Mayans, Romans, and Egyptians, they seem to have folded in on themselves, collapsing beneath the weight of their own civilization. Their demise may also have been caused in part by climatic changes, disease, war, or social unrest. But by the 1400s, the site had been abandoned, and only the mounds remained to speak of their legendary past.

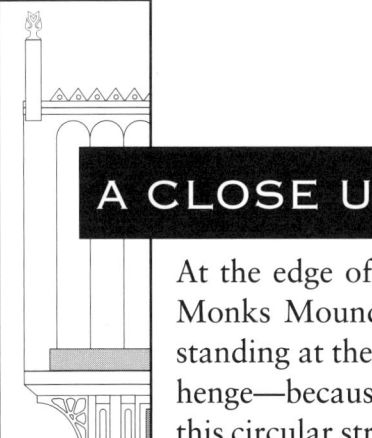

A CLOSE UP

WOODHENGE: CALENDAR AT SUNRISE

At the edge of what was once the city surrounding the great Monks Mound, a series of vertical poles circle a single pole standing at the center. This is the place known today as Woodhenge—because archaeologists believe that Cahokians built this circular structure of wood to serve much the same purpose as the great vertical stones that form Stonehenge in England.

Woodhenge—a reconstructed Cahokian sun calendar similar in function to Stonehenge in England   (Courtesy of Cahokia Mounds State Historic Site)

Here, archaeological excavations have partially uncovered remains of four, and possibly five, circular sun calendars that once consisted of large, evenly spaced log posts. They were probably used to determine the changing seasons and other ceremonial periods of importance to farming. Constructed in about the year 1000, they were an impressive example of Indian astronomical science and engineering.

Today, a reconstructed Woodhenge forms a sunrise horizon calendar consisting of 48 large cedar posts arranged in a 410-foot-diameter circle around a central observation post. Other similar devices marked the seasons and important dates for ancient Cahokians.

Special programs are scheduled at Woodhenge for equinox and solstice sunrises.

PRESERVING IT FOR THE FUTURE

The Cahokia Mounds Museum Society, founded in 1976, is dedicated to promotion and development of the Cahokia Mounds State Historic Site, and the site was designated a World Heritage Site by the United Nations in 1982. It occupies a 2,200-acre tract where the archaeological remnants of the central section of the city known today as Cahokia are located. The museum offers a lecture series, craft classes, and archaeological field schools, as well as seasonal guided tours of some of the site's remaining 80 mounds.

EXPLORING ♦ FURTHER

Books and Articles about Mound Builders

Iseminger, William R. "Mighty Cahokia." *Archaeology*. May/June 1996, 30–37.

Shaffer, Lynda Norene. *Native Americans Before 1492: The Moundbuilding Centers of the Eastern Woodlands*. Sources and Studies in World History series. Armonk, N.Y.: M. E. Sharpe, 1992.

Silverberg, Robert. *The Mound Builders*. Athens, Oh.: Ohio University Press, 1986.

Wood, Marian. *Ancient America*. Cultural Atlas for Young People series. New York: Facts On File, 1990.

Related Places

Hopewell Culture National Historical Park
16062 State Route 104
Chillicothe, OH 45601-8694
(614) 774-1125
Fax: (614) 774-1140

Two thousand years ago the Hopewell Indians of Ohio buried their dead in earthwork mounds here, near the Scioto River. The site includes 23 burial mounds at Mound City Group and large geometric earthworks, including a huge mound in the shape of a serpent. These structures represent a major prehistoric construction effort and offer insights into the social, ceremonial, political, and economic life of the Hopewell people. The finely crafted artifacts of the Hopewell culture (200 B.C. to A.D. 500) found here also show that these highly skilled artisans took part in an extensive trade network east of the Rocky Mountains.

Effigy Mounds National Monument
151 Highway 76
Harpers Ferry, IA 52146
(319) 873-3491

These prehistoric mounds differ from many commonly found from the plains of the Midwest to the Atlantic seaboard—in this general region (and in the Scioto River area of what is now Ohio) they were often constructed in the likeness, or effigy, of mammals, birds, or reptiles. The 1,481-acre area contains 200 mounds, 29 in the shape of effigies. Other mounds are conical, linear, and compound in shape. The mounds were built by Eastern Woodland Indians from 500 B.C. to A.D. 1300.

Ocmulgee National Monument

THOUSANDS OF YEARS ON THE BANKS OF A RIVER
Macon, Georgia

AT A GLANCE

Settlement established: 8000 B.C.

Mounds built: ca. A.D. 900

Fort built: 1690

A memorial to 10,000 years of habitation in this area, from ice age hunters to the Creek Indians of historic times

The park contains a reconstructed 1,000-year-old ceremonial earth lodge, a burial mound, temple mounds, prehistoric trenches, and the site of a colonial British trading post and fort, as well as a visitor center/museum.

Address:
Ocmulgee National Monument
1207 Emery Highway
Macon, GA 31201
(912) 752-8257

> On the Macon Plateau by the Ocmulgee River, numerous civilizations—from the mound builders of the Mississippian period to the Creek Indians of historical times—have interwoven to make this site a multicultural treasure trove.

The Sun is the Supreme Sky Being, the source of food and knowledge, and the Fire is His Mate.
—Woodrow Haney, Sr., Creek Indian

♦ ♦ ♦ ♦ ♦

Great Temple Mound at Ocmulgee rises 45 feet and measures about 300 feet by 270 feet at its base. Lesser Temple Mound, nearby, is similar in form but far smaller. (Courtesy of the National Park Service)

On the bluffs overlooking the muddy, wandering Ocmulgee River, a nomadic people have left behind remains that date back at least 10,000 years, when giant ground sloths roamed the continent, along with other exotic creatures, such as three-toed horses, camels, and mammoths.

At this site, archaeologists have unearthed evidence of six successive civilizations, the earliest possibly dating back as far as 8000 B.C. These people, like those who carved petroglyphs on the West Mesa above the Rio Grande in New Mexico, were probably wandering hunters, descendants of families who crossed the Bering Strait from Asia just before the last of the great ice sheets melted. They moved rapidly across the continent, most likely following game, leaving behind the distinctively grooved Clovis spearheads. For as long as 5,000 years, these prehistoric people lived in what is now Georgia.

Later came the shellfish eaters, groups of people whose main food source, in addition to the deer and bear they hunted, was the mussel beds in the nearby rivers. At Ocmulgee, these people left behind their distinctive type of pottery, made of clay reinforced with fibers of moss or grass. They also left behind spear points they used for hunting and sinkers they used to sink their fishing nets. The shellfish eaters remained until about 100 B.C.

Then, for about the next 800 to 1,000 years, early farmers settled in this fertile countryside—people who, like their predecessors in the region, continued to hunt for food. But they also began planting crops, most likely beans and pumpkins. They too left behind beautiful pottery decorated with elaborate designs. These people, however, were eventually pushed out by another civilization of even more accomplished farmers.

Between A.D. 900 and 1150, an elite society supported by skillful farmers lived on this site near the Ocmulgee River. An outlying segment of the Early Mississippian culture of which Cahokia was a center, these people constructed a town of rectangular wooden buildings, huge pyramidal temple mounds, and at least one burial mound. Circular earth lodges, like the reconstructed earth lodge visitors can explore today, served as places to conduct meetings and ceremonies (see A Close-Up).

They also built a funeral mound here, which visitors can still see today. Originally 30 feet high and more than 250 feet long, it was once a combined temple and burial place and was built in at least seven successive stages, each

Artifacts found on the Macon plateau   (Courtesy of the National Park Service)

stage in a different kind of clay. During excavations in the 1930s, archaeologists uncovered over 100 burials, many of them containing shell and copper ornaments, indicating that this mound was probably reserved for village leaders. The ornaments, tools, clay pipes, and bowls left behind suggest that these people, like the people of Cahokia, believed in an afterlife. Structures built on top of the mound may have been used to prepare the dead for burial. Visitors can still see portions of the Funeral Mound, which unfortunately was partially destroyed during railroad constructions in the 1840s and 1870s.

Of all the monuments built at Ocmulgee, the largest is the Great Temple Mound, originally 40 feet high and 300 feet across at the base, its sides sloping inward, like a pyramid. Built sometime between A.D. 1350 and 1500, this structure's height was achieved laboriously as its builders carried dirt to the site one basketful at a time. At least four different times during this 150-year period, the builders "completed" the mound, capping it off with clay and building a ceremonial structure on top. Then a grander vision took hold and the builders began adding more dirt, enlarging the base and sides and building the great mound yet higher for a perhaps yet grander structure, capped off with a new clay surface. Today, only the grassy mound testifies to this industrious activity, the vision that inspired it, and the grand buildings that served as its crown.

A smaller structure, known as the Lesser Temple Mound, stands nearby. Unfortunately, railroad construction in the 19th century also took its toll on this structure. Another mound, called the Cornfield Mound, was originally about 8 feet high. Beneath it, archaeologists found signs of a cultivated field —a puzzle since Mississippian agricultural fields usually lie in bottomlands, not atop plateaus. The mound itself, according to archaeological evidence, was probably a platform built for a ceremonial building, possibly having to do with harvest rites and ceremonies. Around the east side of the village, two prehistoric trenches have also been discovered, although their purpose isn't certain. Perhaps they were borrow pits, from which dirt for the mounds was carried. Or they may have been used for military defense.

This Mississippian period ceremonial center eventually fell into disuse; archaeologists believe that the earlier group of farmers came back, driving out the group that built the Great Mound. These were ancestors of the Creek (that is, "people of Ocheese Creek," the early name for Ocmulgee River) and the Cherokee. They lived in small villages in the swamps protected from enemy attacks by fences or palisades constructed of upright logs. Known as the Mississippian "Lamar" culture, they also built mounds, and these were the people encountered by the Spanish explorer Hernando De Soto when he traveled through this region in 1540.

The early Creek centered their society on the *idalwa*, or "town," formed by a central settlement and a group of surrounding villages. By the time the Europeans arrived, the Creek had formed a loose confederacy of towns, a nation that incorporated several small tribes into their alliance, including the Alabama, Hitchiti, Yuchi, some Shawnee, and several others, and by the mid 1700s, the Creek Nation consisted of about 60 towns. Town governments were composed of a *micco*, or chief, a subordinate *micco*, and a council. Each town council sent a representative to the General Council, which decided matters of peace and war for the entire nation—although the towns retained a great deal of autonomy.

A town was arranged around a public square, where three main structures were located—a large, round building, called a *chokofa*, where the town

Mound B, Lamar Site—also known as the Spiral Mound; it is the only such mound in existence today. (Courtesy of the National Park Service)

council conducted meetings in winter; a cluster of square public buildings around an open plaza; and a playing field with spectator seating. Family dwellings surrounded the public structures.

The Creek social structure was based on families organized into clans, which sometimes had privileged positions in the nation. Although they did some hunting and gathered nuts and seeds, they also farmed in collective fields or in individual garden plots, raising crops of corn, pumpkins, beans, squash, and sweet potatoes. And, because the Creek were farmers, their religion revolved primarily around medicinal concerns and annual rituals relating to crops, such as the Green Corn Dance, which sometimes lasted four days or more at the end of July or the beginning of August. This symbolic ritual fostered spiritual renewal and celebrated the beginning of the new year that was marked by the ripening of a new corn crop.

But the arrival of the British in this area in 1607 was the harbinger of change. On the coast of Virginia, where they founded the colony of Jamestown, these settlers relied for their livelihood primarily on agriculture—particularly the production and export of tobacco—and at first the impact on the early Creek (as the British called them), or Muscogee (as they called themselves), was small. But as the colony expanded into the interior, the European presence began to pressure the local Indian tribes, pushing them farther and farther inland.

Then yet another factor came to play—and it became a major one. In 1653 the British founded another colony, located much closer to the Macon plateau—Carolina colony—which depended more on fur trade than on agriculture for its economy. Charleston soon became the great fur-trading center of the region, and European traders ventured more than 500 miles westward, as far inland as the Chickasaw villages on the Mississippi River. The impact on the Indians of the Southeast was enormous.

By 1690, the English traders from Charleston, eager to do business with the Creek, built a trading post near the Ocmulgee River. Today visitors to Ocmulgee National Monument can view the site, as well as find extensive interpretation of the Creek experience by park rangers who can offer a wealth of information and materials on the subject.

The new life into which this enterprise drew the Creek played havoc with their stability and their culture. What once was a civilization built primarily on farming now focused more and more on hunting for pelts to trade with the European settlers. They swapped deerskins and furs for firearms, cloth, and trinkets, and they bartered for copper bells, knives, guns, and rum. In addition to deerskins, they began to engage in Indian slave trade for the Caribbean sugar plantations. The Creek society they once had built began to degenerate rapidly as the Creek way of life changed forever.

Eventually, in 1715, as they began to feel pushed out of their territory, the Creek took a stand against the white invaders—with Emperor Brim of the Creek launching what became known as the Yamasee War (named after one of the tribes) to drive out the British, as well as the French and Spanish, all of whom had made inroads on the Carolina territory of the Creek. The war lasted until 1717, ending in defeat for the Creek and abandonment of the village at Ocmulgee.

The story became uglier as tensions increased between European settlers and Native Americans. Georgia was founded in 1733 between South Carolina and Spanish Florida, with colonist James Oglethorpe negotiating with resident Creek Indians for the right to settle his colonists at the mouth of the

Savannah River. From this base, European settlements expanded nonstop across lands claimed by the Creek and Cherokee.

The story of the descendants of the Ocmulgee mound builders continued, however. Farther west, the French had begun to expand their settlements along the Mississippi, scattering and engulfing many tribes, including the Natchez. By this time, English settlers on the southern coast began to become alarmed by what they saw as French encroachments on their own area of expansion. They began to try to enlist the help of the Creek and Cherokee, as well as the Chickasaw and the Choctaw to help them stop the French. The Cherokee remained aloof and the Choctaw sided with the French, but the British obtained cooperation from both the Creek and the Chickasaw. All four of these groups, which represented powerful confederations of tribes from the southeastern regions, traded heavily with European traders. Some intermarried and assimilated European ways.

The Cherokee, however, endured great pressure from the Europeans and, despite efforts to fight the tide, were forced to give up large expanses of territory in 1755 and 1770. During the Revolutionary War, most Cherokee, as well as Creek, sided with the English, which unfortunately also didn't help their cause after the war. Continued interference from Spanish and British agents, in an effort to block the success of the newly formed government, fomented unrest and, although the new federal government attempted to make treaties with the Indians, in reality most European settlers and land speculators showed complete disrespect for the Indians' rights. The Creek were forced to cede huge areas of their territory in Georgia—regions their people had held for many centuries. Border warfare broke out, and as the War of 1812 began between the United States and Britain, many angered Creek responded to appeals by the Shawnee chief Tecumseh to stand fast, joining with other tribes in a massive resistance against white advances.

Ultimately, however, they met defeat. Finally, with a series of treaties spanning the years from 1773 to 1832, the Creek ceded lands to the incoming settlers in a devastating series of events. In 1832, Indian delegates signed a treaty giving up part of their lands in Alabama in exchange for 320 acres for each Indian family, and 640 acres for each chief. The families could stay on their land or sell their lots and move west at government expense to lands where autonomy was promised them. The setup invited deceit and violence. Fraudulent land agents tricked Creek land recipients out of their holdings,

Creek houses were burned, and families were forced from their homes. Angered Creek retaliated by killing white settlers and destroying their homes, barns, and crops. Finally, in 1836, some 2,500 Creek were rounded up, marched to Montgomery, crowded onto barges, and sent down the Alabama River on the first leg of a forced removal to a hot, arid area in what was known as Indian Territory (now Oklahoma).

During the summer and winter of 1836 to early 1837, more than 14,000 Creek were forced to make the three-month journey to Oklahoma, traveling over 800 land miles and another 400 by water, with little more than the clothes they were wearing. Decimated by European diseases such as small-

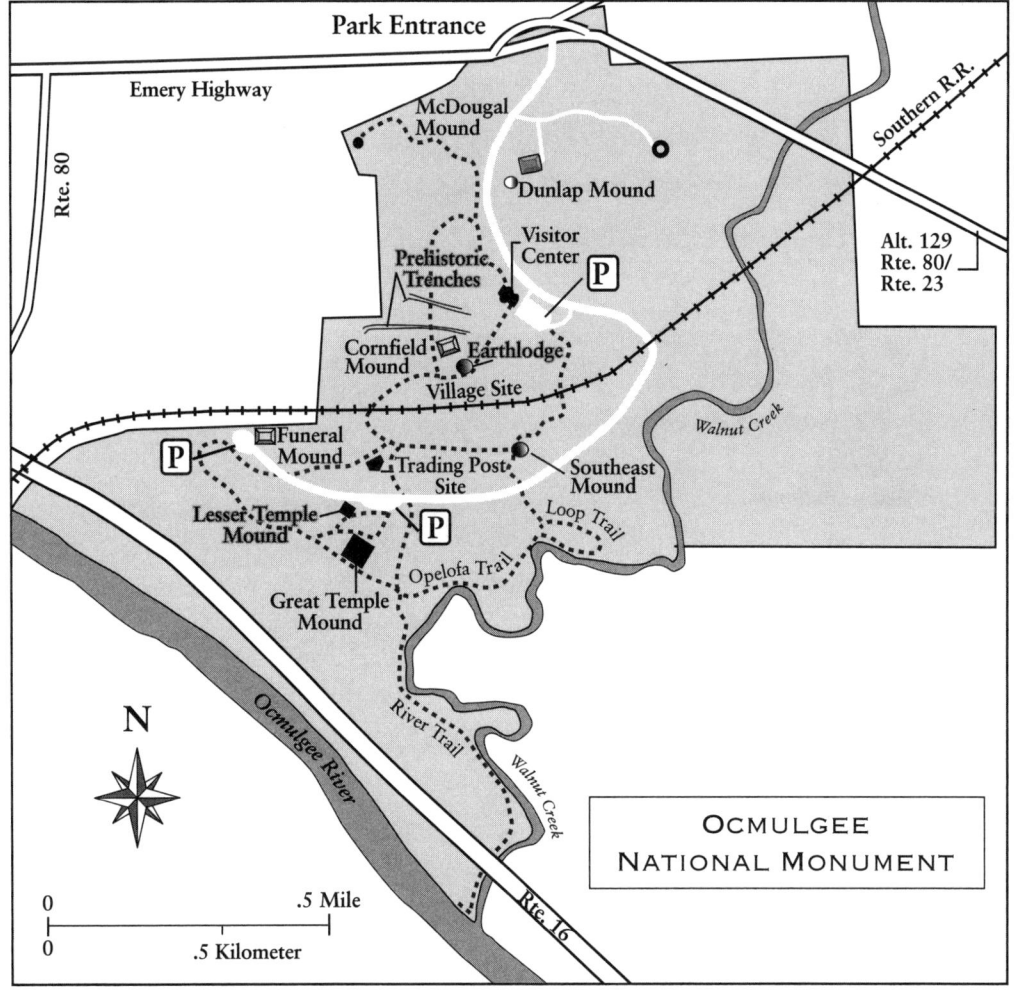

pox, against which they had no immunity, frostbitten and exhausted, more than 3,500 Creek died on the road that became known as the Trail of Tears. Survivors arrived at their destinations ill prepared to fend for themselves, sick, and in pitiful condition, and many died shortly afterward. Yet, while the once powerful Creek Nation had suffered greatly, its spirit was not destroyed. Despite continued hardships, its citizens carved a new life for themselves in Oklahoma, their descendants remaining a proud and sovereign people, one of the largest Indian nations in the country, numbering 43,550 in 1990.

Today, the mounds of Ocmulgee remain as a reminder of a culture that once was rooted here for centuries. When naturalist William Bartram journeyed through Ocmulgee in the 1770s, he described the mounds he saw as the "wonderful remains of the power and grandeur of the ancients in this part of America." They still evoke a large civilization rooted in a time now far removed, and they remind us of the story of members of the Creek Nation who later resided here.

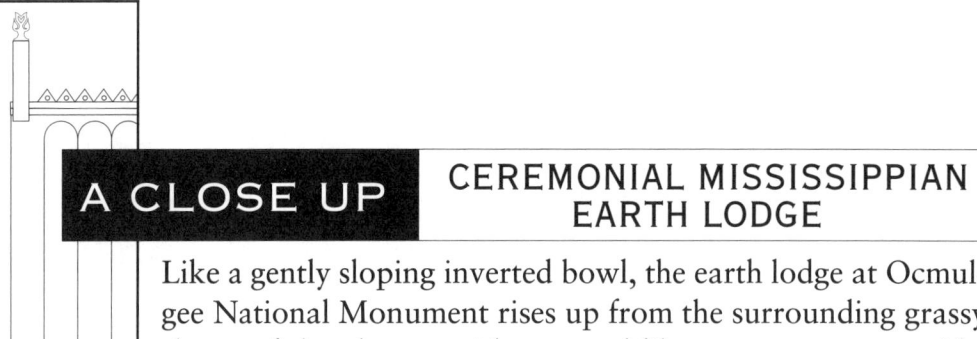

A CLOSE UP — CEREMONIAL MISSISSIPPIAN EARTH LODGE

Like a gently sloping inverted bowl, the earth lodge at Ocmulgee National Monument rises up from the surrounding grassy slopes of the plateau with a tunnel-like entrance on one side, braced by timbers. Here, visitors can stand inside an authentic replica of one of the earliest public buildings in North America, a council chamber of the mound-building Muskogean Indians.

On the site where an authentic earth lodge council chamber of the Middle Mississippi culture once stood, this reconstruction was built upon the original clay floor, which is now about a thousand years old. Located on the north side of the Mississippian village, this lodge was probably a meeting place for the town's political and religious leaders. In the original, destroyed by fire in about A.D. 1000, only one vessel remained, suggesting that the mound builders destroyed the earth lodge on purpose, leaving nothing behind to be learned by invaders.

Archaeologists have determined that the ceiling made of cane would have been first to succumb to the heat of the fire, letting the roof of thick clay fall through to the ground and reducing the intensity of the fire. The burning roof timbers then would have collapsed, resembling the charred spokes of a huge wheel resting inside the giant clay tub that was once the room of the earth lodge. But although the original building was gone, the fire left behind a sort of charcoal blueprint from which archaeologists could reconstruct the earth lodge with confidence.

The mound builders brought the clay from the valley of the Ocmulgee River, paddled and smoothed it into place, packed it into a firm foundation by walking on it, and then allowed the sun to bake it. Then they built the earth lodge above the clay floor. Here, during the winter season, amidst the smell of smoke and earth, the master farmers gathered within this ceremonial earth lodge where respected men and head warriors from many villages came to discuss the political, religious, and material affairs of their region.

In ancient times, the passageway into the earth lodge was twice as long as it was high, preventing the cold winds from entering and the elders' words from escaping to the ears of eavesdroppers. The doorway was placed so that twice a year, on February 22 and October 22, the sun could stream through this passageway and bathe the high priest in light. The significance of these two days, however, we do not know.

We do know, though, that fire played an important part in the ceremonies held here by the master farmers. A gray layer of ash was left by fires that burned in the baked, clay-lined hearth located exactly in the center of the circular room. Possibly these fires represented sacred fire brought with the Muskogean tribes when they migrated east from across the Mississippi and invaded the Macon plateau. Additionally, the mounds and the symbol found here of a forked eye on the head of a bird suggest an influence from Central America.

The high priest sat on a center seat positioned so that his line of sight formed a center line across the fire pit, due east and directly out through the entrance tunnel. The supporting posts formed a perfect 18-foot square, with the east-west beams parallel to the high priest's line of sight. The chamber itself resembles an amphitheater, with 47 seats that become wide and rise slightly, one above the other on the circular clay bench, the higher and wider seats possibly reserved for those who were older and more important than others who attended the meetings.

At the foot of each seat a scooped-out basin in the floor may have held ceremonial items or may have been connected with the only artifact left behind by the fire, a clay pot, which may have contained a black drink made from the dried leaves of cassena holly mixed with willow bark and button snakeroot. This substance caused vomiting and was passed by the men from one to another before each ceremony in a conch shell. Believing the process cleansed and purified them, one by one they sipped the black drink and vomited, possibly into containers for vomit that they kept in the depression at their feet.

In recognition of the master farmers' knowledge that the sun is the ultimate source of all our energy, Woodrow Haney, Sr., a Creek Indian, has commemorated the spirit of these ceremonies with the following words, which are recorded in the park's audiovisual program:

> The Sun has the power of life and death. He looks down upon the earth, and as long as He keeps his flaming eye upon us, we are safe. If this Eye of Creation is turned away, we will die.
> The Sun helps us succeed in war, and shows us the broad path. A path that leads to victory and brings us safely home.
> Our sacred Fire represents the Great Sun on Earth, for they are partners.
> Our Fire is wise, and acts in concert with the Sun. The Old People say, "If you do anything wrong, in the presence of the Fire, the Fire will tell the Sun, before you can run the length of one arm."
> The Sun is the Supreme Sky Being, the source of food and knowledge, and the Fire is His Mate.

Each spring evening during the Cherry Blossom Festival in March, as the blossoms burst on the cherry trees in Macon, the park conducts lantern-light tours of the park, which culminate at the top of the Great Temple Mound. From there you can look down on the luminary-lit earth lodge and imagine another time, when an ancient people met there at fire-lit gatherings to settle the issues of their community and cleanse their spirits.

PRESERVING IT FOR THE FUTURE

Known as the most scientifically excavated of the South's major Indian sites, Ocmulgee National Monument is recognized as one of the most important and best interpreted examples of mound sites in the Atlantic coast region. It preserves nine ceremonial mounds, a funeral mound, and a restored ceremonial earth lodge of the Early Mississippian period. Visitors can drive along a 2-mile road to gain easy access to several of the earthen mounds, including the Great Temple Mound (the largest of the park's seven mounds) and the Funeral Mound. Five miles of trails also connect the major features of the park.

The Visitor Center houses a major archaeological museum, including exhibits illustrating the human habitation of the area from 10,000 B.C. to the early 1700s, with emphasis on the Mississippian town that flourished here from A.D. 900 to 1100. Visitors can also see a short film, *People of the Macon Plateau*, and visit the museum's Native American Art Gallery.

EXPLORING ♦ FURTHER

Books and Articles about Ocmulgee, the Creek and the Cherokee

Green, Michael D. *Creeks*. Indians of North America. New York: Chelsea House, 1990.

Hally, David J., ed. *Ocmulgee Archaeology*. Athens: University of Georgia Press, 1994.

Mankiller, Wilma. *A Chief and Her People: An Autobiography by the Principal Chief of the Cherokee Nation*. New York: St. Martin's Press, 1993.

Norman, Geoffrey. "The Cherokee: Two Nations, One People," with photos by Maggie Steben. *National Geographic*, May 1995, pp. 72–ff.

Shemie, Bonnie. *Mounds of Earth and Shell: Native Sites: The Southeast Native Dwellings*. New York: Tundra Books, 1993.

Woodward, Grace Steele. *The Cherokees*. Norman: University of Oklahoma Press, 1963.

Related Place

Horseshoe Bend National Military Park
11288 Horseshoe Bend Road
Daviston, AL 36256
(205) 234-7111

For General Andrew Jackson, the battle against the Creek Confederacy at a horseshoe-shaped bend in the Tallapoosa River was a key victory in the War of 1812. It provided him with lustrous publicity, set him up for victory against the British at New Orleans, and started him on his path to the presidency. But to the Indian warriors he defeated and their families, the loss was devastating. It was the culmination of a clash of cultures known as the Creek War of 1813–14 coinciding with the war against the British; it set Indian against Indian and Indian against white American.

The troops Jackson led into battle on March 27, 1814 had combined men from the 39th U.S. Infantry and the Tennessee Militia with allies from the Cherokee and Upper Creek nations. Their foes—warriors from the Red Stick Creek (also known as the Lower Creek)—chose this horseshoe bend as their last stand against the European invasion into their lands and for the preservation of the Creek Confederacy. From behind a wooden rampart built across the peninsula formed by the river bend, the Red Stick defended their position valiantly. But they could not hold it. More than 800 Creek died at Horseshoe Bend, the largest number of deaths suffered by Native Americans in any battle against U.S. troops. Most of their lands were lost, forcing the Creek westward out of Georgia and Alabama. This park of 2,040 acres preserves the battle site and provides interpretation of the conflict of cultures that took place here.

Mesa Verde National Park

CLIFF DWELLINGS OF THE MESA-TOP FARMERS
Mesa Verde National Park, Colorado

AT A GLANCE

First settlements built: ca. 600

Cliff dwellings built: 1150–1300

Home to the Northern San Juan Anasazi in what is now Colorado

Ancient pit house excavations, cliff dwellings, and pueblo ruins built by the peoples known as the Anasazi, probable ancestors of several modern Pueblo groups, such as the Hopi, Zuni, and Tewa.

Address:
Mesa Verde National Park
Mesa Verde National Park, CO 81330
(970) 529-4461 or 529-4465

> *Visitors to these cool, wooded mesa tops and sheltered side canyons quickly understand what attracted the Ancient Ones to this place as early as A.D. 600. Here, between A.D. 1150 and 1300, during the period now known as the Great Pueblo, the Anasazi built beautiful, massive sandstone structures. Of all the cliff dwellings left by the Anasazi, those at Mesa Verde are among the most impressive.*

> . . . the Cliff Dwellers of Mesa Verde numbered only three thousand but lived in some of the most stunning and enduring buildings of all time, built in the recesses of the cliff walls.
> —Alvin M. Josephy, Jr. in *500 Nations*

♦ ♦ ♦ ♦ ♦

Mesa Verde—Spanish for "Green Plateau"—offers an unparalleled opportunity to see and experience the life of the "Ancient Ones," the Anasazi. During the summer months, visitors can walk through numerous mesa-top villages and the five spectacular major cliff dwellings, built between 600 and 1300 by the people the Navajo called *Anasazi*—literally, "ancient enemies." It's a strange name to use, as pointed out by the Hopi and the Zuni, who descend from the Anasazi. But the people who lived here left no record of what they may have called themselves, and the Pueblo peoples who descend from them use many different names. So, while *ancestors of the Pueblo peoples* might be a preferable term, it's a little awkward, and most people still refer to these Ancient Ones as Anasazi.

These first inhabitants were basket weavers, who perched their homes on top of this green plateau dotted with woodlands of piñon-juniper, ponderosa, and yellow pine. They built simple pit houses, which they created by scooping out dirt to create a shallow, sunken floor—either round in shape or square with rounded corners—and an antechamber, with the entryway always oriented toward the south. In the center was the hearth and ash pits, and wing walls sometimes divided living areas inside the house. A roof built of branches and twigs was supported by vertical timbers, with wooden lean-to walls supported in turn by the roof, and a covering of mud over the whole structure. Visitors can see a reconstruction at Mesa Verde that gives a sense of how small, cramped, and dark these homes must have been, although providing relatively good protection against the cold and heat of the arid southwestern climate.

The basket maker peoples were fairly mobile foragers, but they also cultivated beans and corn, and they ground up seeds and grain with small stones called manos and metates. They gathered in small communities, and their housing was never elaborate—usually only about 15 feet in diameter—but they sometimes kept small dogs for pets, sentries, and perhaps pack animals. They made and decorated large conical baskets, wore jewelry made of shells and stones, and sandals woven of yucca fiber.

Within 200 years, the people had moved out of their pit houses and had begun building structures above ground in small villages. By now, to their diet of beans and corn they had added squash and melons, and they hunted with bows and arrows. The mesas made an ideal farming location, with mild winters, hot summers, lengthy growing seasons, dependable rains in late summer, and fertile soils. To protect against drought, they also built small run-off dams and reservoirs to collect water.

The park abounds with wildlife —it contains three wilderness areas —and you may catch a fleeting glimpse of a mountain lion, bobcat, deer, mountain sheep, or elk, much as the basket makers often did when they lived here. You may also see eagles and hawks or even peregrine falcons as they soar over the park's

Early inhabitants of the mesa tops built pit houses like this one, dating from ca. A.D. 550. (Courtesy of the National Park Service)

Most kivas, like this one in Step House at Mesa Verde, were roofed by "bricklaying" logs to form a dome. In this photograph, you can see the fire pit and the deflector behind it, as well as the pilasters and banquette. Behind the deflector is the ventilator shaft. (Courtesy of the National Park Service)

north rim. The plant life is also startling, a unique blend of species found in the mountains and deserts. It's easy to imagine why this might have been an attractive place to live, even long, long ago.

By 1150, as many as 5,000 people lived on the mesa top. By now, the simple structures of one story were gone, replaced by multistory structures —sprawling apartment houses composed of 200 or 300 rooms. The people who lived here had also begun to reach out in their influence, to outlying communities as far away as 100 miles. During this time there was plenty of food, and life was comfortable. People had time to be creative—weaving and making pottery with intricate shapes. But within 100 years this golden age came abruptly to an end, concluded by some sudden or cataclysmic change, such as a major change in climate or an attacking enemy or internal strife. At this point, the Mesa Verde Anasazi left the mesa tops for the safety of the caves that nature had spectacularly carved in the cliffs below.

Cliff Palace, the largest cliff dwelling in North America, contains 217 rooms and 23 kivas. It once was home to 200 to 250 people. (Courtesy of the National Park Service)

These are the dwellings for which Mesa Verde is famous—huge complexes of multistory buildings with hundreds of rooms and dozens of kivas, or ceremonial meeting places, built into the magnificent towering cliffs below the mesa. The cliffs alone leave one in awe, and the dwellings convey a sense of master builders at work. What was it like to live perched on the side of a cliff, tucked halfway into a cave?

Only about 3,000 people made the move from the top of the mesa, and it must certainly have been difficult, requiring each of the millions of building stones that made up these walls to be carried into the canyon. No water was nearby, so that had to be carried in, too. Travel to and from the farms on the surface above was probably accomplished by climbing the walls using toeholds and finger holds. But the place was safe, not easily attacked, and protected from inclement weather.

Even this move didn't last long, however. By 1275 or 1276 the rain stopped falling almost altogether. Many of the Anasazi who had lived here moved away, and they never came back, even though the rains began again in about 1300.

The first photograph to record these spectacular structures was taken by William Henry Jackson in 1874, but the vast size of the complex of structures was not recognized until 1888, when Richard Wetherill and Charles Mason came upon what is now known as Spruce Tree House (see A Close-Up). In addition to this structure, which is open year-round, visitors can see Balcony House, Cliff Palace, Long House, and several others (although not all these are open year-round and some require a strenuous hike). You reach Balcony House by climbing a 32-foot ladder and crawling through an ancient crawl space, which really gives you the feel for what it may have been like to live in this place. Cliff Palace, with its remarkable number of large circular kivas, was probably a center where rituals and ceremonies were performed for inhabitants of some of the small outlying communities in the region.

Although Mesa Verde's peak building years came to an end by 1275, known as the end of the time of the Great Pueblo, the people who dispersed from there joined other communities and carried with them the culture and

Balcony House, one of the three largest cliff ruins at Mesa Verde   (Courtesy of the National Park Service)

the skills that had made them the master builders of the San Juan Anasazi—
the Ancient Ones who lived among the San Juan Mountains.

A CLOSE UP — SPRUCE TREE HOUSE

Spruce Tree House, the third largest cliff dwelling (after Cliff Palace and Long House) among the several hundred located within the boundaries of Mesa Verde National Park, was constructed by the Anasazi between A.D. 1200 and 1276. Its 114 rooms and 8 kivas were built into a natural cave that runs

Spruce Tree House at Mesa Verde National Park   (Courtesy of the National Park Service)

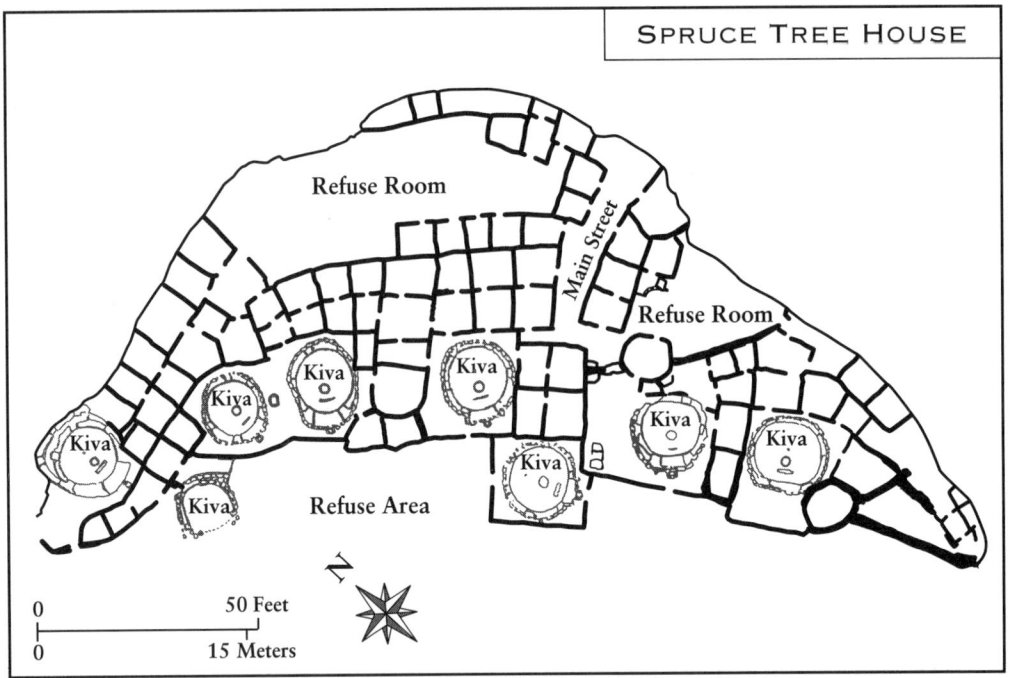

As many as 100 to 125 people once lived in the 114 rooms or apartments of Spruce Tree House and met with their neighbors in the sunken kivas for spiritual convocations and community gatherings. During the day the courtyards buzzed with activity—children playing, women making pottery or grinding corn, and men making tools. Unwanted items were thrown down the hill into garbage heaps—which have provided many clues about food, pottery, and tools used by those who lived here.

a maximum of 89 feet deep and 216 feet wide. Probably about 100 people lived here. The overhanging cliff has protected Spruce Tree House from weather deterioration, and it remains in very good shape.

First discovered in 1888 by two local ranchers, the dwelling had a large spruce tree (and so the name) growing from in front of it to the top of the mesa. The ranchers used the branches of the tree to reach the dwelling, but the tree was afterward cut down by another explorer.

The first courtyard at Spruce Tree House was probably noisy with activity when people lived here—as women ground corn into flour or made pottery, men made stone tools or blankets from turkey feathers or cotton, or prepared for the summer planting. Older people would have been talking, children playing, and the place ringing with laughter and

The first courtyard at Spruce Tree House (Courtesy of the National Park Service)

the sounds of dogs barking and domesticated turkeys gobbling. Ladders from the courtyard lead down into two kivas (their roofs have been restored by the National Park Service). T-shaped doors lead into the rooms in which people lived. When the weather was cold, animal skins or sandstone slabs were used to cover the doorways.

Because Spruce Tree House has suffered so little deterioration, about 90 percent of its structure is original and intact. But the Anasazi who lived here had used up much of the surrounding forests to build their homes, had depleted the land with farming, and had exhausted much of the small game in the area. The prolonged drought from about 1276 to 1299 didn't help, either. The Anasazi began migrating south during the late 13th and early 14th centuries—shortly after they built Spruce Tree House—and their descendants, the Hopi and other pueblo dwellers, now live in northern New Mexico and Arizona.

PRESERVING IT FOR THE FUTURE

Established by Congress on September 29, 1906, Mesa Verde is the first national park set aside to preserve the works of humans. Mesa Verde was also designated as a World Heritage Cultural Park on September 8, 1978 by the United Nations Educational, Scientific and Cultural Organization (UNESCO), which preserves and protects both the cultural and natural heritage of designated international sites.

You can reach most of the ruins through self-guided hikes, although some are steep. Some ruins are available for viewing only on ranger-led hikes. Good views also can be seen from canyon rims opposite the sites, and bus tours are available from about May through mid-September.

More information is available at the Far View Visitor's Center and Chapin Mesa Museum. The museum itself contains a large number of artifacts, outlining not only what archaeologists found in digging at Mesa Verde but also how they know what they know about the people who inhabited the mesas. Exhibits explore dendrochronology (establishing dates by analyzing tree rings), pottery dating by styles and methods, and many other fascinating aspects of the science of archaeology.

EXPLORING ♦ FURTHER

Books about Mesa Verde and the Anasazi

Arnold, Caroline, and Richard Hewitt. *The Ancient Cliff Dwellers of Mesa Verde*. New York: Clarion Books, 1992.

Cordell, Linda S. *Ancient Pueblo Peoples*. Edited by Jeremy A. Sabloff. Washington, D.C.: Smithsonian Books, 1994.

Liptak, Karen. *Indians of the Southwest*. The First Americans Series. New York: Facts On File, 1991.

Mays, Buddy. *Ancient Cities of the Southwest: A Practical Guide to the Major Prehistoric Ruins of Arizona, New Mexico, Utah, and Colorado*. With

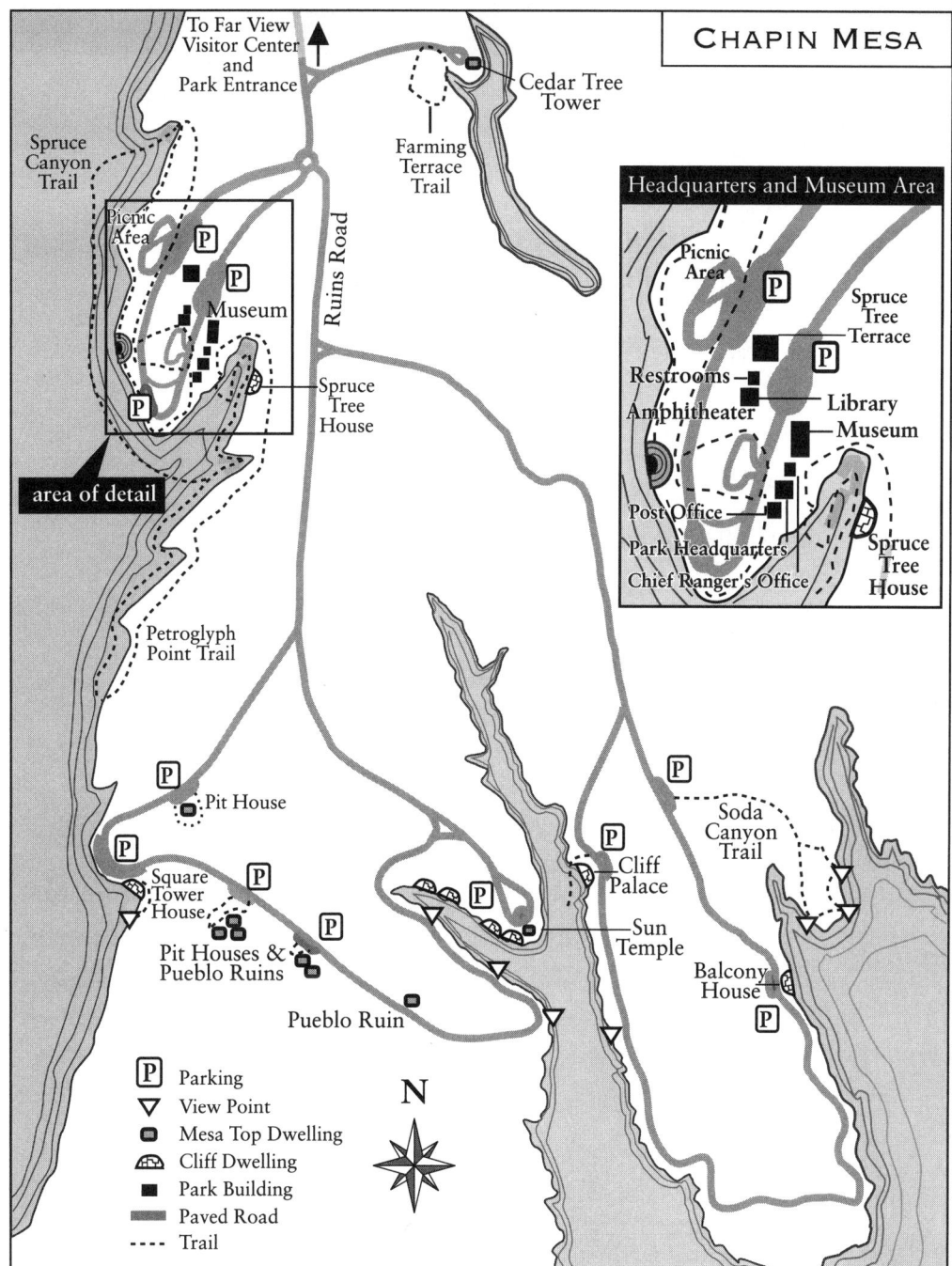

Map of Chapin Mesa, one area of Mesa Verde. Many of the most spectacular sights at Mesa Verde can be seen from lookouts along the roads atop Chapin Mesa. Sites range from pit houses and pueblo ruins on the mesa top to views across the canyon of impressive cliff dwellings, including Cliff Palace.

MESA VERDE NATIONAL PARK

photographs by Buddy Mays and foreword by Joseph C. Rumberg, Jr. San Francisco: Chronicle Books, 1982.

Petersen, David. *Mesa Verde National Park*. New True Books. Chicago: Children's Press, 1992.

———. *The Anasazi*. New True Books. Chicago: Children's Press, 1991.

Rohn, Arthur H. *Mug House, Mesa Verde National Park—Colorado*. National Park Service Archeological Research Series—7D. Washington, D.C.: U.S. Government Printing Office, 1971.

Related Places

Bureau of Land Management Anasazi Heritage Center
27501 Highway 184
Dolores, CO 81323
(970) 882-4811

The Anasazi Heritage Center houses approximately 2 million records, samples, and artifacts found during the creation of the McPhee Dam and Reservoir. Its exhibits display many objects found and encourage visitor involvement. Here visitors find dozens of activities with interactive computers, microscopes, corn-grinding implements, a loom and other weaving materials, and "touch me" exhibits. A holographic image, orientation film, archaeological test trench profile, and partially reconstructed full-size pit house also help bring the process of archaeological investigation to life.

Dominguez and Escalante Ruins
Next to the Anasazi Heritage Center
27501 Highway 184
Dolores, CO 81323
(970) 882-4811

In 1975–76, the United States Bureau of Land Management sponsored the excavation of the two 12th-century sites known as the Dominguez and Escalante Ruins. They were named for two Franciscan priests from Santa Fe —Fathers Dominguez and Escalante—who in 1776 recorded their discovery of an ancient ruin in this area.

Escalante Ruin, which may be the settlement the two priests found, was a Chaco outlier, a satellite community of the Chaco Canyon urban center—despite its proximity to the Northern San Juan Anasazi region. Built in 1129, possibly as a trading post, the pueblo sits atop a small hill with a spectacular view of the Montezuma valley and the whole Four Corners area.

Dominguez Ruin, discovered in modern times, is nestled in a small depression just south of the Anasazi Heritage Center. Built to house one or, at most, two families, the Dominguez site was probably linked to the larger Escalante complex, although its structure indicates its builders were probably San Juan Anasazi.

Lowry Ruins, excavated in the 1930s, and then restored in the 1960s, can be reached via a self-guided trail. Lowry (named for homesteader George Lowry) is representative of the medium-sized pueblos sprinkled throughout the Montezuma valley. This pueblo probably housed about 100 people at its peak in the early 11th century, when it included about 40 rooms and 8 kivas. It was abandoned about 1150. One of the more remarkable features of Lowry is the adjacent Great Kiva. In 1967, the site was designated a National Historic Landmark. Unlike most other sites, Lowry exhibits at least two different Anasazi cultural traditions—using styles similar to the Chaco Anasazi from the south as well as patterns from the Northern San Juan Anasazi, such as those found, for example, at Mesa Verde.

Chaco Culture National Historical Park

GREAT PUEBLO CITY OF THE SOUTHWEST
Nageezi, New Mexico

AT A GLANCE

Built: Mid-800s

Home of Chaco Culture Anasazi from the 850s to ca. 1150

An Anasazi community in Chaco Canyon, probably the largest in the Southwest, scattered up and down the 25-mile canyon cut into the desert of northwest New Mexico. Principal ruins: Pueblo Bonito, Great Kiva of Casa Rinconada, Chetro Ketl, Una Vida, Hungo Pavi, Kin Kletso, Casa Chaquita, Pueblo del Arroyo.

Address:
Chaco Culture National Historical Park
P.O. Box 220
Nageezi, NM 87037-0220
(505) 786-7014

> At its peak, the Chaco Canyon community was, according to archaeologists, the New York City of the ancient Southwest. But today, as you turn off the highway at Nageezi to drive toward Chaco Canyon, the landscape seems vast and lonely.

The district is little better than a desert: many parts of it,
indeed, are absolutely barren wastes of sand and rock
which do not even support the usual dry-country flora
of the Southwest.
—Alfred V. Kidder, archaeologist, 1924

♦ ♦ ♦ ♦ ♦

Pueblo Bonito was built in stages from approximately 850 to the mid-1100s. (Dave Six. Courtesy of the National Park Service)

CHACO CULTURE NATIONAL HISTORICAL PARK

The dry, hot wind of Chaco Canyon whips unceasingly across the gray sand, wild grass, lonely shrubs, and piles of sandstone and shale. Its name means "desert," and without question it seems an unlikely spot for perhaps the largest ancient community in the Southwest, a great dwelling place built by the people known as the Anasazi. But here the ancestors of the modern Pueblo peoples of New Mexico and Arizona built 14 elegant multistoried stone towns during the period between 850 and 1150.

Like those who carved drawings in the rocks along the escarpment and plateaus of the West Mesa on the Rio Grande, these are descendants of the earliest travelers who came across the Bering Strait land bridge and followed game southward toward warmer weather. The Anasazi settled and lived for about 2,000 years in the region where the borders of Utah, Arizona, Colorado, and New Mexico meet, known today as Four Corners, and their territory extended south nearly to the headwaters of the Colorado River in Arizona and covered most of northern New Mexico. The second major group of pueblo dwellers were the Mogollon, who settled southern New Mexico and northern Mexico.

By A.D. 600 Chaco Canyon was occupied by large groups of Basket Maker Anasazi. These farmers used the runoff from occasional desert storms to irrigate their crops of corn, and they lived in pit houses, similar to their Mesa Verde neighbors to the north. By the 800s, they had moved from the pit houses into the first pueblos—compact, one-story structures that they built clustered close together near springs or other sources of water.

The population of the Chaco area grew rapidly, however. The river then was not dry, as it is now, and this steady water supply and nearby farmland attracted settlers. Within a century and a half the canyon experienced a great housing boom, with many new pueblos being built. Some were very large. Pueblo Bonito, which may have housed as many as a thousand people, was probably the single largest building in the Southwest at the time it was built, rising to at least four stories, with some 600 rooms and 33 kivas (ceremonial chambers), as well as examples of extremely fine masonry.

Without the help of metal tools, the Chaco Anasazi constructed these enormous communal buildings with admirable craftsmanship. Evolving their methods over the centuries, they developed from simple walls that were only one stone thick, visible in the oldest walls in Pueblo Bonito. Then as they sought to build higher, more extensive walls, they learned how to place thick

Kin Kletso was built in two stages, the first dating from about 1125 and the second from 1130 or later. This pueblo, which may have risen three stories high on one side, had about 100 rooms and 5 enclosed kivas. (Dave Six. Courtesy of the National Park Service)

inner cores of rubble inside a thin veneer of facing stone, tapering the wall as it rose. These constructions involved considerable planning in their most complex forms, during the Classic period (1020–1120). In Kin Kletso, another type of wall is visible, built with a thin inner core of rubble and thick outer veneers of shaped sandstone, similar to masonry found in the Mesa Verde region at about the same time.

Chetro Ketl was also one of the largest Chaco structures, begun about 1020 and completed around 1054, with further remodeling about 50 years later. The Great Kiva and three elevated kivas may indicate that this pueblo had special religious significance, designed for large gatherings. In an extraordinary discovery, several wooden objects, including birds, prayer sticks, arrows and discs, were found here in the Great Kiva, probably used in ceremonies. Chetro Ketl's enclosed plaza, which is clearly visible, is typical of great houses of this period.

Both Pueblo del Arroyo and Kin Kletso were apparently built in stages. The central part of Pueblo del Arroyo was begun about 1075, with additions made in 1101 of the north wing and in 1105 of the south wing. Smaller than the others, it had 280 rooms, but more than 20 kivas. And Kin Kletso dates from 1125, with a second stage added five or more years later. Much smaller, Kin Kletso rose three stories high but had only 100 rooms, housing maybe 200 or 300 people, with five enclosed kivas.

We know from the artifacts found in the regions of the Four Corners area that both differences and similarities existed among the peoples of this time

Chetro Ketl—begun in 1020, it was completed by 1054 and then remodeled and enlarged in the early 1100s. The entire complex held about 500 rooms and 16 kivas. (Dave Six. Courtesy of the National Park Service)

Pueblo del Arroyo was built in stages over a relatively short time, from around 1075 to 1110.
(Dave Six. Courtesy of the National Park Service)

and place—in the way they built structures, designed pottery and basketry, drew pictographs and petroglyphs, and farmed. The Chacoans' architecture and pottery, possibly even their language and dress, were different from the Northern San Juan Anasazi, in Mesa Verde. In contrast to the agricultural and populous northern group, the Chaco Canyon dwellers developed a relatively complex, more sophisticated society. By A.D. 1000, the Chacoans were comparatively advanced in architectural styles and construction.

An irrigation system was constructed in the Chaco Wash near the central towns, which became the site of extensive agricultural enterprises. But the Chaco people clearly also relied on their extensive trade routes, as evidenced by a 300-mile network of roads that linked this central point on the Chaco Canyon with some 400 settlements, known to archaeologists as Chaco outliers, which were spread throughout the Four Corners area, including the

San Juan Basin of Colorado, Utah, and New Mexico. These roads extend north and south from a dozen or so Chacoan central towns to other Chaco villages. The Chacoans used this system of outliers to supplement food supplies at home, trading ornaments and other items they had finely crafted for food and other raw materials. Archaeologists have found evidence of extensive trading with far-away contacts as well, including the Coastal Indians and Mesoamericans, the earliest residents of Guatemala, Belize, Honduras, and Nicaragua. The Anasazi had parrots from Central America as well as pottery from nearby communities specializing in clay objects. They also brought in turquoise and other stones cut from far-off quarries, from which they fashioned exquisite ornaments and jewelry. Clearly, they were able to communicate and trade with a wide variety of peoples. They also may have exercised considerable religious and political influence over other Anasazi groups.

Today the river that once ran through the broad, shallow Chaco Canyon has run dry, and the people have long-since left. But thirteen ruins remain, many of which stood four to five stories high, with beautifully finished walls and some eight hundred rooms. The Anasazi built two main types of dwellings—cliff dwellings, found in the mouths of caves carved out of the many cliffs along the canyon walls; and free-standing dwellings known as pueblos, often having many rooms and multiple stories. Among the Anasazi ruins, we also find kivas (circular semi-underground ceremonial rooms) and sometimes a Great Kiva (very large rooms for public ceremonies) and towers.

The major Chaco towns are large, four- and five-storied structures with elaborate Great Kivas, like the one at Casa Rinconada (see A Close-Up). It's hard to imagine this desolate, silent desert area so filled with activity and commerce. Yet research shows that it was once a bustling, vibrant community. Future research will very likely yield even more interesting facts, at which these silent stone walls for the moment can only hint.

A CLOSE UP: CASA RINCONADA—THE GRAND KIVA

Unlike the other Great Kivas at Chaco Canyon—which are enclosed within plazas of the great houses such as Pueblo Bonito and Chetro Ketl—the Great Kiva at Casa Rinconada stands alone, away from the house structures. Elevated slightly, it was otherwise much like the other Great Kivas: It had a raised firebox, a low bench (or banquette) around the perimeter, wall niches for storing special ceremonial objects (such as turquoise, shell beads, and prayer sticks), a subfloor passage for entering and leaving the kiva, and

Casa Rinconada, located on the canyon's south side, is the largest Great Kiva in Chaco Canyon. (Dave Six. Courtesy of the National Park Service)

an antechamber. Most likely, this type of isolated kiva provided a meeting place for small villages in the vicinity.

Archaeologists have inferred the use of the Anasazi kivas from the uses that present-day Pueblo peoples make of their own kivas. Among modern Pueblos, groups of men and women form societies to care for the village's spiritual needs. One society, for example, might have responsibility for conducting ceremonies and rituals for ensuring a favorable growing season. Another might take charge of curing illness, or guaranteeing successful hunts, or inducing harmony in the village. Ceremonies of this type are the primary function of the kiva, but when not in use in this manner, the community also often uses it as a place for social gatherings or as a work area. Presumably, similar uses were made of the Great Kiva of Casa Rinconada and the other kivas of Chaco Canyon.

PRESERVING IT FOR THE FUTURE

The first Europeans to record their observations in Chaco Canyon arrived in 1850. Several major excavations took place from that time into the early 20th century. On March 11, 1907 the site was proclaimed Chaco Canyon National Monument and became a national park December 19, 1980, administered by the National Park Service. It was designated a World Heritage Site December 8, 1987. Having an elevation of 6,100 feet, the park includes 33,989 acres of land.

In traveling to Chaco Canyon, be prepared for stretches of unpaved roads, often impassable in wet weather. A road leading north from the Visitor's Center takes you to all major excavated ruins in Chaco Canyon. Self-guiding trails (none of them strenuous) explore seven of the park's ruins, including Pueblo Bonito, Chetro Ketl, Pueblo del Arroyo, Casa Rinconada, and three village sites. With back country permits, four other day-hiking trails are also available. In addition, rangers offer guided tours as well as evening camp fire programs. Visitors are implored (and required by federal law) to leave all objects of antiquity and ruins as they find them.

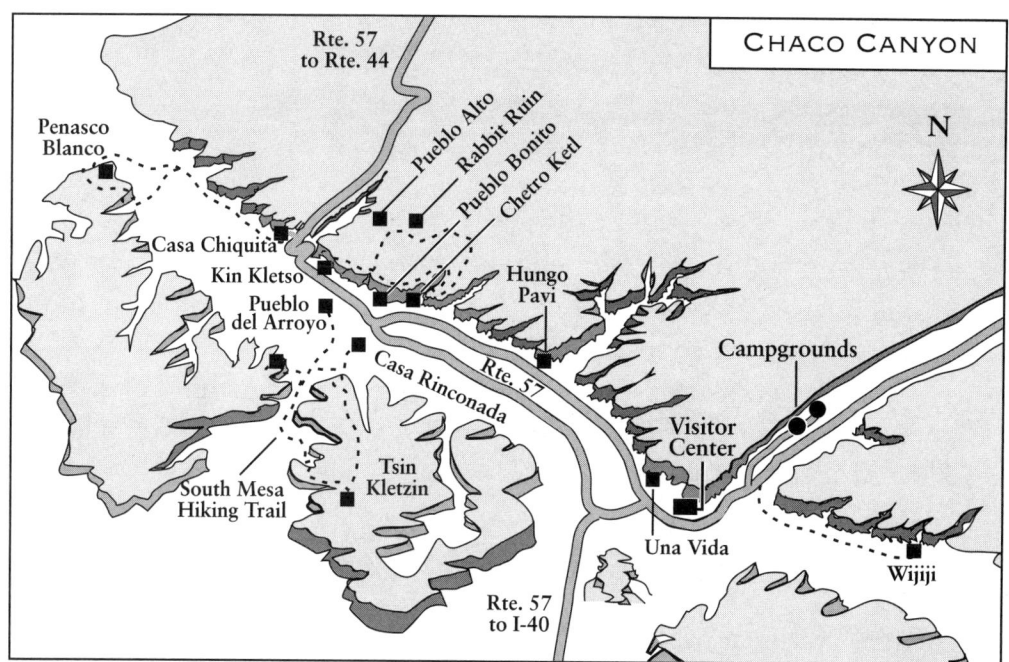

The sights at Chaco Canyon National Park range from the pueblos built along the canyon bottom to high mesa dwellings.

EXPLORING ♦ FURTHER

Books about Chaco Canyon and Its Inhabitants

Burby, Liza N. *The Pueblo Indians*. The Junior Library of American Indians. New York: Chelsea House, 1994.

Cordell, Linda S. *Ancient Pueblo Peoples*. Edited by Jeremy A. Sabloff. Washington, D.C.: Smithsonian Books, 1994.

Frazier, Kendrick. *People of Chaco: A Canyon and Its Culture*. New York: W.W. Norton & Co., 1988.

Gabriel, Kathryn. *Roads to Center Place, A Cultural Atlas of Chaco Canyon and the Anasazi*. Boulder: Johnson Books, 1991.

Judge, W. James. "Chaco Canyon—San Juan Basin." In *Dynamics of Southwest Prehistory*, edited by Linda S. Cordell and George J. Gumerman 209–261. Washington, D.C.: Smithsonian Institution Press, 1989.

Lekson, Stephen H. *Great Pueblo Architecture of Chaco Canyon, New Mexico*. Publications in Archaeology, 18B, Chaco Canyon Studies. Albuquerque, N.M.: National Park Service, USDI, 1984.

Liptak, Karen. *Indians of the Southwest*. The First Americans Series. New York: Facts On File, 1991.

Mays, Buddy. *Ancient Cities of the Southwest: A Practical Guide to the Major Prehistoric Ruins of Arizona, New Mexico, Utah, and Colorado*. With photographs by Buddy Mays and foreword by Joseph C. Rumberg, Jr. San Francisco: Chronicle Books, 1982.

Powell, Susan. *The Pueblos*. A First Book. New York: Franklin Watts, 1994.

Vivian, R. Gwinn. *The Chacoan Prehistory of the San Juan Basin*. San Diego, Calif.: Academic Press, Inc., 1990.

Related Places

Aztec Ruins National Monument
P.O. Box 640
84 County Road 2900
Aztec, NM 87410-0640
(505) 334-6174
Fax: (505) 334-6372
E-mail: azru interpretation@nps.gov
Internet: http://www.nps.gov/azru

This is one of the most significant prehistoric archaeological sites affiliated with the Chaco and Mesa Verde Anasazi or Ancestral Pueblo cultures. Visitors can walk through a series of rooms with original roofs (constructed beginning about 1108) and a Great Kiva reconstructed in 1934. Established in 1923 as a National Monument, Aztec Ruins were designated a World Heritage Site in 1987.

El Morro National Monument
Route 2, Box 43
Ramah, NM 87321
(505) 783-4226

In addition to "Inscription Rock," which bears hundreds of inscriptions from the historical era, visitors can also view pre-Columbian petroglyphs and

Pueblo Indian ruins at this site, which was proclaimed a national monument on December 8, 1906. Visitors can observe inscriptions and fragile petroglyph carvings on a ½-mile hike past the base of the cliff. A second trail, the Mesa Top Trail, is a 2-mile loop passing over the top of the mesa past ancient pueblo ruins.

Escalante Ruins
Next to the Anasazi Heritage Center
27501 Highway 184
Dolores, CO 81323
(303) 882-4811

Built in 1129, this Chaco outlier was probably a trading post, located closer to the Northern San Juan Anasazi settlements of Mesa Verde, Lowry, and Hovenweep than to Chaco Canyon 100 miles to the south. Nearby Dominguez ruins housed only one or two families and were built in the Northern San Juan style. (See Related Places at the end of Mesa Verde National Park.)

Knife River Indian Villages National Historic Site

EARTH LODGE DWELLERS ON THE PLAINS
Stanton, North Dakota

AT A GLANCE

Built: ca. 1100

Reconstruction period: 1830s

Remains of villages of Plains tribes, dating from ca. 1100 to 1837

This national park protects the site of the largest and most sophisticated villages of the Plains Indians in this area, including the Hidatsa, Mandan, and Arikara tribes. Remains of terraced fields, fortifications, and earth lodges, including a reconstructed earth lodge like the ones that were built here. Home of Sacagawea, the Indian guide who helped explorers Lewis and Clark cross the Great Divide.

Address:
Knife River Indian Villages National Historic Park
P.O. Box 9
Stanton, ND 58571-0009
(701) 745-3309

> Positioned on a tributary to the Missouri River, amidst the rolling, grassy plains of North Dakota, Knife River Villages were poised for greatness in the 18th and early 19th century.

Four Bears never saw the white man hungry, but when he gave him to eat . . . and how have they repaid it! . . . I do not fear death . . . but to die with my face rotten, that even the wolves will shrink . . . at seeing me, and say to themselves, that is Four Bears, the friend of the whites.

—Four Bears, Mandan chief,
as he died of smallpox, ca. 1837

These circles of raised earth with central depressions are all that remains of a large, once lively town that flourished along the riverbank. (Fred Armstrong, Courtesy of the National Park Service)

KNIFE RIVER INDIAN VILLAGES NATIONAL HISTORIC SITE

The Mandan, closely intertwined with the neighboring Prairie Hidatsa and Arikara, were among the earliest Plains peoples, having migrated westward by 1400 from the Ohio River and Great Lakes country. They first stopped along the upper Missouri River in territory now part of South Dakota, later moving farther north to settle near the mouth of the Heart River, now part of North Dakota. When European Americans first made contact with them in 1700, they had settled near the Big Bend of the Missouri, living in permanent villages, like the Hidatsa to the north and the Arikara to

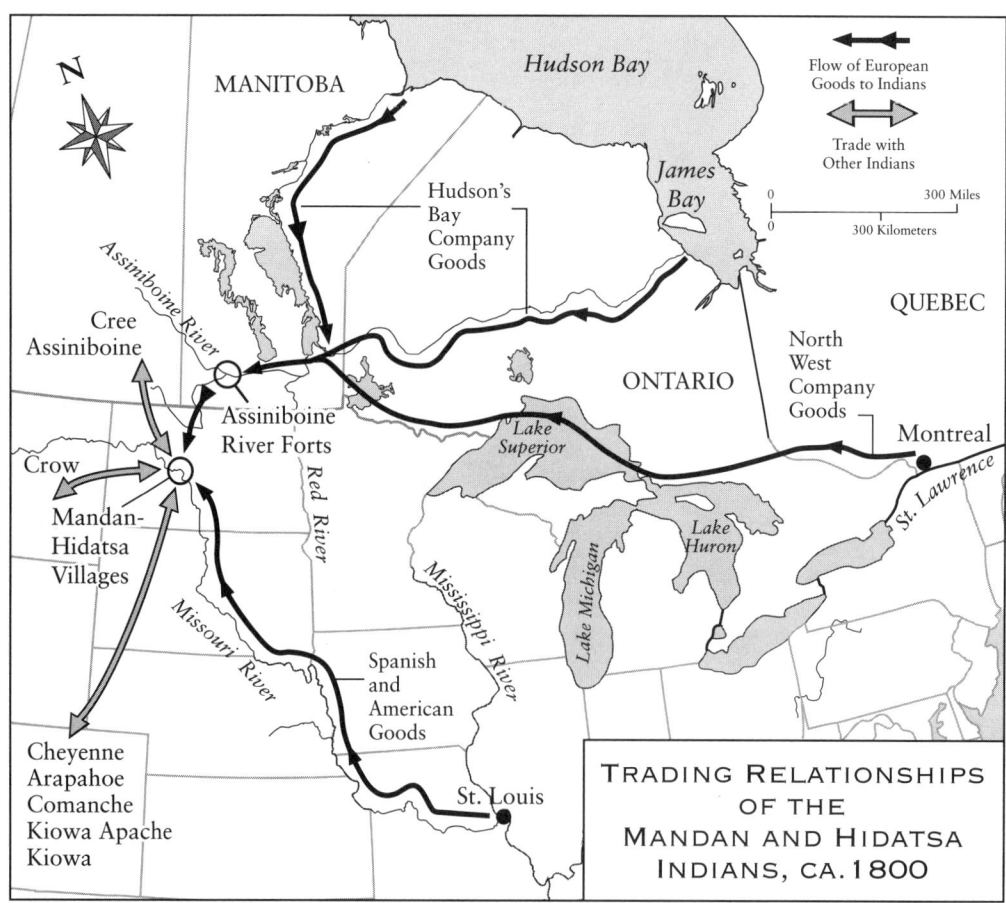

The Mandan and Hidatsa were centrally located for trade on the upper Missouri River—on everyone's way up or down the river. Their villages became great trade centers for the exchange of goods. (Map adapted from Stanley A. Ahler, Thomas D. Thiessen, and Michael K. Trimble, *People of the Willows: The Prehistory and Early History of the Hidatsa Indians*. Grand Forks: University of North Dakota Press, 1991; redrawn by Jeremy Eagle)

This full-sized, reconstructed earth lodge, completely furnished with items that would have been found there in 1830, provides a realistic experience of what it might have been like to have lived in one of these dwellings. (Bill Lutz, Courtesy of the National Park Service)

the south. There they cultivated a variety of crops, including corn, squash, beans, sunflowers, and tobacco. At least once a year, they ventured into the grasslands to hunt buffalo, and after the mid-1700s, when they had horses, they roamed even farther in pursuit of the great herds.

In 1738, the first white explorers came through—a family of explorers named Verendrye, from Quebec. Living on the upper Missouri, the Mandan, Hidatsa, and Arikara were on everyone's way up or down river, and the Three Affiliated Tribes, as they are sometimes called, in their location at what is now the Knife River Villages site, had long maintained an important trading center for many native peoples. Nomadic hunters had come there to barter meat and hides for crops for many years. With contact with European Americans and Canadians, more goods began to flow through their trade channels, including items from the Hudson's Bay Company to the north, as well as obsidian from Yellowstone Park; shells from the Pacific, Atlantic, and Gulf coasts; and copper from the Great Lakes. For these, they exchanged flint from their own region.

Other whites passed through their town—Meriwether Lewis and William Clark stopped by in 1804 (where they took on Sacagawea and her husband, the French-Canadian trapper Charbonneau, as guides). Painter/writer George Catlin stayed with the Mandan during the period when he was traveling, from 1830 to 1836, among many native peoples, and he wrote about the Mandan extensively and painted several famous portraits of Mandan warriors and Hidatsa villages. Also, the Swiss painter Karl Bodmer passed through, capturing numerous images of Mandan life on canvas.

But the open friendliness of the Mandan and Hidatsa toward their white visitors backfired. With the white explorers came their diseases, against which the Indians had never had the opportunity to build up immunity, having never before been exposed. In 1837, a devastating smallpox epidemic hit the villages, and of an estimated 1,600 Indians living in this area, only 125 survived.

Today, as one stands on the grassy knolls above the Missouri River, the deep depressions where Hidatsa earth lodges once stood are all that remain of what once was a thriving trade center. By 1845, all the Mandan and Hidatsa had left, and the Arikara tribe followed in 1862.

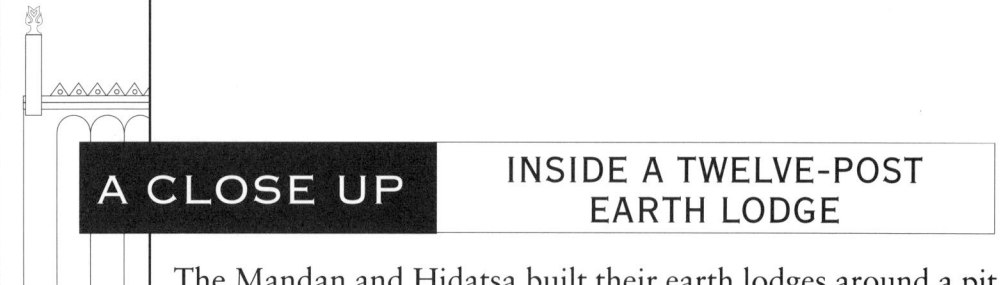

A CLOSE UP — INSIDE A TWELVE-POST EARTH LODGE

The Mandan and Hidatsa built their earth lodges around a pit dug 1 to 4 feet deep, where they set up four interior supporting posts, near the center, hammered into the ground by builders and joined by four horizontal beams. On the exterior perimeter, 12 supporting posts were likewise sunk into the ground, joined horizontally by stringers around the top. Next, leaners and rafters were added to provide support for the walls, and finally, the builders added wall coverings made of willows, grass, and earth. The roof of the Hidatsa village earth lodge was domed and covered with earth. In Mandan villages, people often climbed onto their roofs, where they gathered to gossip, doze in the sun, or do chores.

Inside, beds were positioned around the perimeter beneath the stringers, leaving free of clutter the space between the circumference of the earth lodge

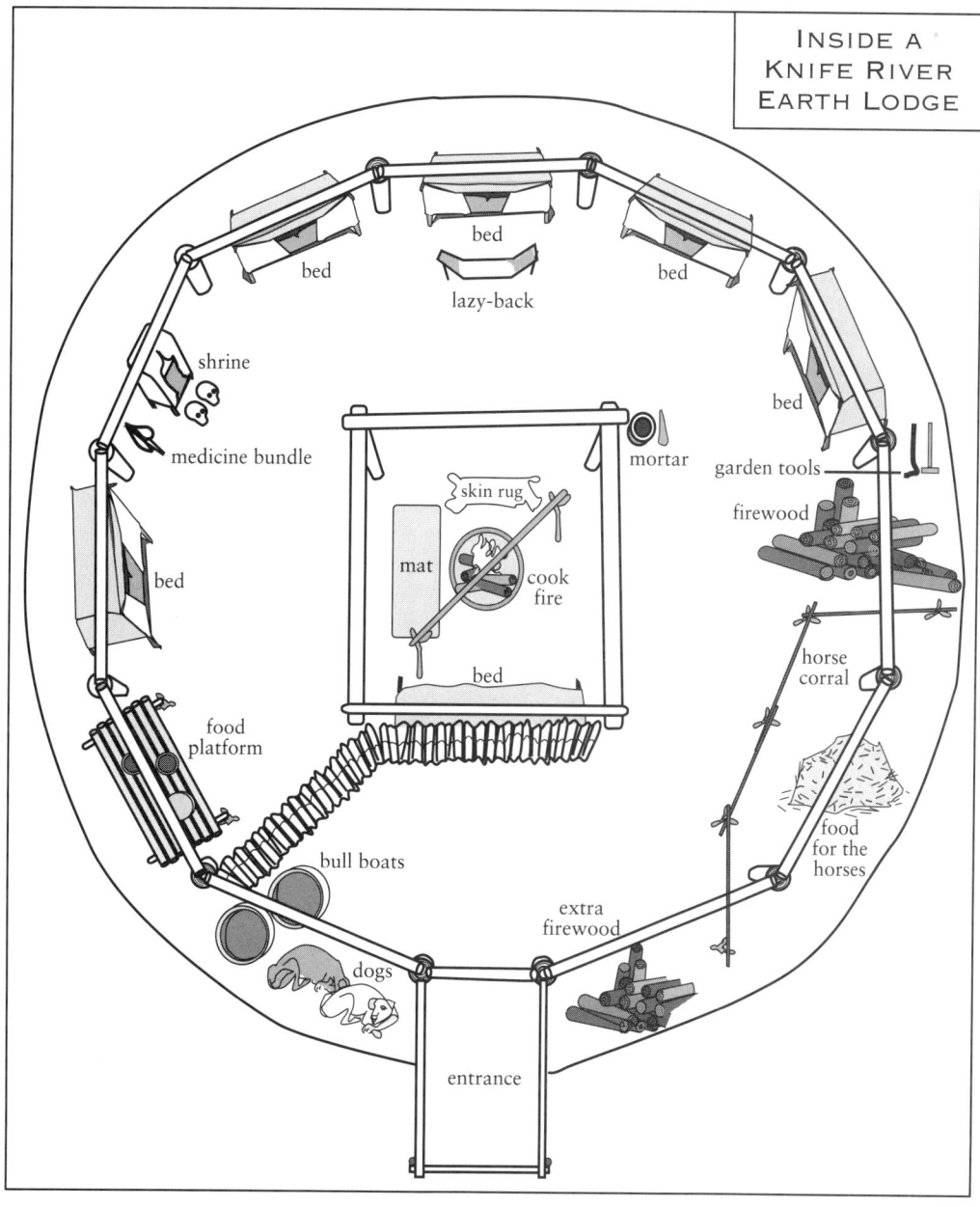

and the center hearth, which had a smoke hole above it. Personal belongings, firewood, a medicine bundle, gardening tools, dogs, and even a small corral for horses, accessible from the outside, might also be spaced out along the edges of the walls. A food platform was arranged near the door, possibly partitioned off. Bullboats, used for hauling meat and hides across the Missouri after buffalo hunts, might be piled against one pole.

In the center was the hearth, traversed by a long pole for cooking and drying meats and skins. Here space was set aside for the cook. And a special place of respect by the fire would be saved, where an elder might sit to make arrows or rest.

PRESERVING IT FOR THE FUTURE

A National Historic Site was established at Knife River Indian Villages in 1974 to preserve remnants of the culture and agricultural ways of life of the

Buffalo Bird Woman (Grace Henry), whose ancestors once lived in the Mandan and Hidatsa villages on the Missouri River banks, 1991 (Fred Armstrong. Courtesy of the National Park Service)

Northern Plains Indians. It is the only national historic site dedicated to the preservation of the prehistory and history of the Plains Indians.

EXPLORING FURTHER

Books about the Mandan, Hidatsa, and Other Plains Indians

Ahler, Stanley A., et al. *People of the Willows.* Grand Forks, N.D.: University of North Dakota Press, 1991.

Catlin, George. *Letters and Notes on the Manners, Customs, and Conditions of North American Indians.* Vol. 1. New York: Dover Publications, 1973.

Freedman, Russell. *Buffalo Hunt.* New York: Holiday House, 1988.

Hoxie, Frederick E. *The Crow.* Indians of North America. New York: Chelsea House, 1989.

Lepthien, Emilie U. *The Mandans.* New True Books. Chicago: Children's Press, 1990.

Schneider, Mary Jane. *The Hidatsa.* Indians of North America. New York: Chelsea House, 1989.

Related Place

Buffalo Bill Historical Center
Museum of the Plains Indians
720 Sheridan Avenue
P.O. Box 1000
Cody, WY 82414
(307) 587-4771 or (800) 227-8483
Fax: (307) 587-5714

The Plains Indian Museum (established in 1969) features an extensive collection of Plains Indian art and artifacts. Its exhibits and interpretative programs explore the cultural histories of the Plains Indian peoples, including the Arapaho, Crow, Cheyenne, Kiowa, Comanche, Blackfeet, Lakota, Gros Ventre, Shoshone, and Pawnee, from buffalo-hunting days to the living traditions of the present.

Sitka National Historical Park

WHERETLINGIT MET RUSSIAN
Sitka, Alaska

AT A GLANCE

Occupied by the Tlingit: ca. 1650–1804

1804 battlefield on which members of the Tlingit clan Kiksadi resisted invasion by Russian traders

The site of an Indian fort where Tlingit warriors tried to hold off an attack from Alexander Baranov, head of the Russian-American Company. Also an outdoor museum featuring authentic Tlingit and Haida totem poles.

Address:
Sitka National Historical Park
P.O. Box 738
106 Metlakatla
Sitka, AK 99835
(907) 747-6281

> In this place Katlian, leader of the Tlingit, faced down the formidable Russians in ships and Aleut warriors in dugout canoes who came to attack and take the land from his people. The Tlingit called these messengers from the far-off tsar the Iron People.

When the Iron People, the Russians, came to Alaska in vessels much larger than canoes, they had weapons that smoked and made noises like thunder. On their vessels they had larger weapons that hurled balls of iron that would smash trees into pieces.

—Tlingit legend of Katlian and the Iron People, compiled by Dee Alexander Brown

The culture of the Tlingit people is powerfully evoked by the carved figures they created, such as this house post portraying Mother Eagle. (Courtesy of the National Park Service)

SITKA NATIONAL HISTORICAL PARK

Practically from the time the Russians had first arrived in 1799 in the Alaskan panhandle, relations had gone badly between them and the Tlingit Kiksadi clan. The Russians came to trap furs, and they soon established the Russian-American Company. The Tlingit knew much that could help the Russians—such as survival and hunting skills—but the Kiksadi quickly saw that cooperating meant answering to the faraway but powerful tsar of Russia. It meant they would have to work for free for the fur company. It meant the Russians would interfere in their lives and dealings. In exchange what would they gain? European goods such as tobacco, sugar, and firearms—attractive, but no exchange for independence and freedom. Hostility was present from the beginning, and outright violence broke out in 1802, when Tlingit warriors attacked Redoubt St. Michael, a Russian outpost several miles northwest of Shee Atika on what is now Baranof Island. Nearly everyone at the settlement was killed in the attack—both Russians and Aleut, many of whom, having a different cultural background from the Tlingit, had allied themselves with the Russians. The Russians withdrew, but not for good.

Katlian, leader of the Tlingit of Shee Atika, was a man of sinewy strength and unbendable will. Shrewdly, he knew when the Russians withdrew in 1802 that they would be back, and he prepared his warriors. They built a sturdy wooden fort east of their village. And they armed for battle with breastplates of wood, many wearing helmets carved in the image of fierce animals and giant warriors. Katlian himself wore a helmet in the shape of the raven, a powerful symbol to his people, and wielded a blacksmith's hammer tied to his wrist.

Late in September 1804, the Russians returned in their great sailing ships. They demanded that the Kiksadi surrender their village and leave. Instead, the warriors fortified themselves within their fort. Bombardment from the Russian gunboat *Neva* and three other ships began October 1. But the fort held firm, sustaining little damage. The Russians and their Aleut allies stormed the fort and were bloodily sent running. The following siege lasted for six days and nights. On the seventh day when the Russians approached the fort, they found it empty. The Kiksadi had finally run out of flints and gunpowder and during the night they had silently fled. The Russians lost no time building a fortified settlement on the spot where the Tlingit home of many generations had once stood.

Katlian led the Tlingit warriors in battle against the Russians and Aleut in 1804 at this battleground. The Tlingit succeeded at first in repulsing the attackers. But only a few days later, overcome, they had to surrender their stronghold—and with it the village their clan had called home for generations. (Courtesy of the National Park Service)

Finally, 17 years later, in 1821, the Russians invited the Tlingit back to help with the trapping and hunting. From that time on, until the Russians finally withdrew from Alaska in 1867, an uneasy truce existed between the two factions. And the Russians were safe only as long as they were vigilant.

The Tlingit resisted resolutely because they had learned that living in this land shaped by glaciers and fiercely cold winters required tenacity above all. These quiet fjords and peaceful islands of hemlock and spruce were their home, shared with the Haida and Tsimshian Indians. They had learned how to maneuver the waters in canoes—some as long as 60 feet—which they fashioned from cedar trunks by burning, steaming, and carving the wood. Having little farmland, they turned to the ocean for their food, using spears and hooks to snatch cod, halibut, and herring from the icy waters. In summer they fashioned traps to catch the salmon as they swam upstream to spawn in the shallow, rocky rivers.

With their neighbors, they had divided the region up so that each Tlingit clan had exclusive fishing waters; infringement by others was grounds for harsh retribution or even war. Each group had its own, closely guarded, secret trade routes along the coast and to the interior. Skillful bargainers met with representatives from other tribes. They sought items unobtainable on their part of the coast, such as caribou skins, fox furs, jade, and copper. For these they traded dried fish, otter furs, and highly valued woven Chilkat robes, an art form that only master designers and weavers could produce. However, with the arrival of the new invaders—first the Russian fur traders, followed by prospectors, lumberjacks, and fishermen, all seeking wealth from Tlingit lands—the carefully made agreements regulating hunting grounds and trading trails no longer held the same significance.

Totem poles such as this raven crest pole can be viewed along the trail that winds from the Visitor Center to the site of the Tlingit Fort. (Courtesy of the National Park Service)

The Tlingit were a sociable, outgoing, and generous people, among themselves. They gathered for many kinds of occasions, celebrating births and weddings, mourning deaths, and performing dances prior to hunting and fishing expeditions. When the long, dark hours of the northern winter loomed, they celebrated the traditional arrival of that season with an impressive ceremony known as the potlatch, an unusual display of conspicuous consumption. The host of a potlatch invited guests to his home for a period often lasting several days and distributed his possessions among them. This lavish gesture of hospitality left the host divested of material goods, but rich in social standing.

The Tlingit had developed a culture rich in symbolism and artistic expression—much of it ceremonial

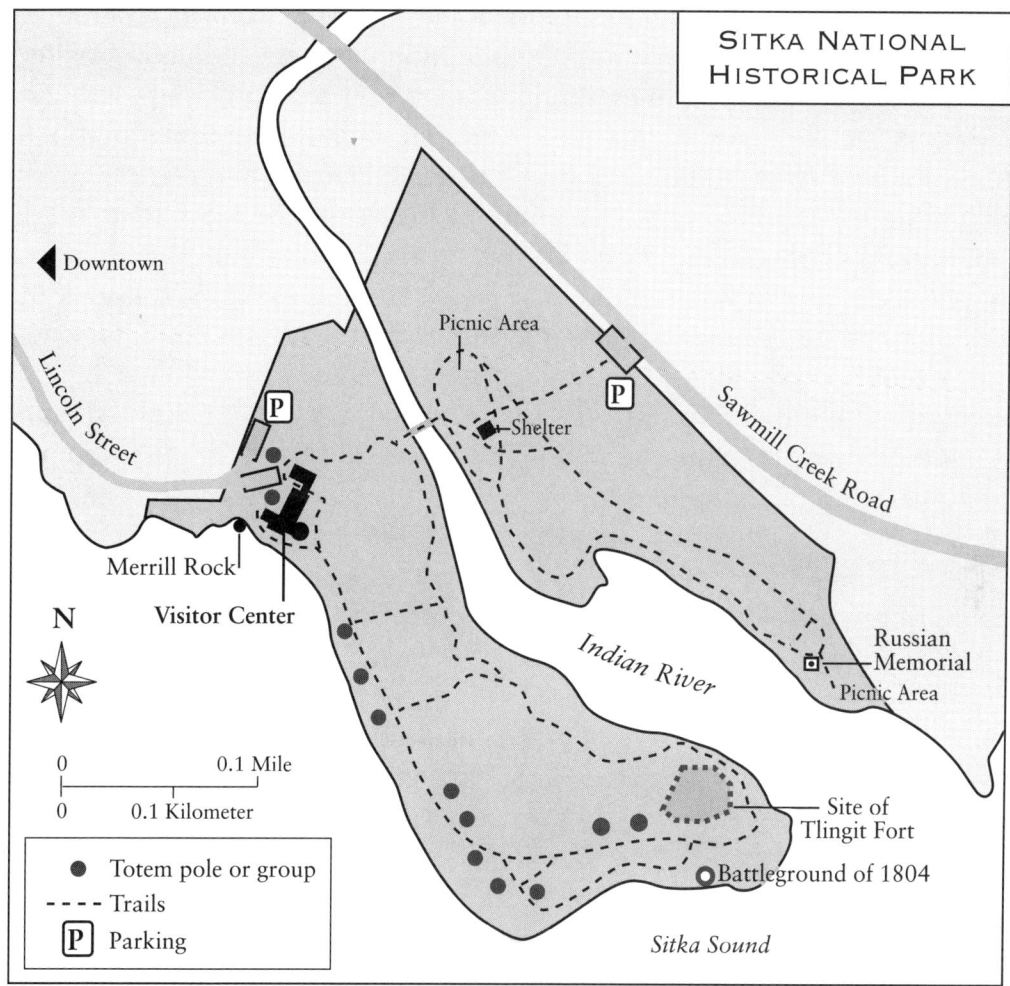

The main unit of Sitka National Historical Park commemorates the Battle of Sitka, which took place in 1804 between the native Tlingit and Russian fur traders. It is located on a spit land at the mouth of Indian River on Baranof Island in the Alexander Archipelago in southeast Alaska.

and shamanic, including carved staffs, masks, and rattles, decorated for potlatch songs and dances, as well as other rituals. But the culmination of their skills was the figures, or totems, carved on great standing poles of cedar. (The Tlingit traded with the Haida to the south, where the prized large cedar trees grew.) Comparable to family crests, these figures told a story or legend or evoked an event or tradition, usually symbolized by the shape of an animal, such as a bear, eagle, killer whale, or raven. The Tlingit had no written language, and the stories were handed down orally from generation to generation. Many have been lost or altered, sometimes reflecting the original intent, and sometimes not. Often, the symbols had an evocative power that brought to the Tlingit's mind a host of related stories, ideas, and emotions, much in the way the cross does for Christians. People placed totem poles, often bearing the family crest, in front of their houses. They used them to tell the legends of the clan. They carved them as memorials to a beloved chief or to celebrate an event such as a child's birth or a hunter's brave deed. Occasionally, they were used to elicit shame or exact payment of a debt, when someone had wronged the clan or village.

But when Presbyterian missionaries followed the wave of fortune-seekers to southeast Alaska in the late 1800s, they were shocked by these enormous monuments in cedar—some as tall as 90 feet—carved in the abstract likeness of animal and human forms. They misunderstood their significance, thinking the poles represented some form of idolatry, and many of these irreplaceable works of art and cultural artifacts were destroyed.

More than a dozen totem poles are preserved, however, in Sitka National Historical Park. And, although these totem poles were not there when Katlian led the Kiksadi against the Russians and Aleut, some similar to them certainly were, and the poles' carvings seem to speak through the generations to something universal in us all. Today you can walk the trail—now lined with totem poles—to the end of the spit of land jutting out into Sitka Sound and see the spot where the Kiksadi built their fort and held off the Iron People for six triumphant days some two centuries ago. In 1971, as a result of settlements of claims against the government, the Alaska tribes recognized as the Tlingit received nearly $1 billion as well as millions of acres of land. With these funds a tribal corporation was established. In 1990, there were 13,925 Tlingit living in the United States.

A CLOSE UP: THE CHILKAT ROBE

Traditionally, among the Tlingit, both making and owning a Chilkat robe or blanket was an honor reserved for the wealthy. These unique, handwoven garments of mountain goat wool played an important part in ceremonial occasions —often worn to potlatches, where they might be presented to honored guests. Or they might be worn in ceremonial dancing, or to wrap a body lying in state.

Colored dyes of yellow, dark brown, and greenish blue were originally made from natural sources—a lichen called wolf moss for the yellow,

Chilkat robes like this modern example form part of the Tlingit traditional ceremonial dress, along with carved masks and stylized weapons. (Courtesy of the National Park Service)

hemlock bark and boiled urine for the brown, and copper boiled in urine for the green-blue. But by the 1890s, commercial yarns and dyes came into use, although the traditional colors continue to prevail.

Creating the robes was a product of teamwork. The patterns, designed by men on pattern boards, were highly stylized, making use of clan symbols and natural forms in an abstract geometric decoration. Eyes were often used as fillers and animals laid out flat. As in other art forms, the designs on the robes were open to interpretation, sometimes telling a story, although the designer alone knew his actual intent. The weaving was done by women, from wool gathered by hand from goats then carded and rolled into thread by hand, using a strip of cedar bark rolled into the thread for the warp. Then, seated before a simple hand loom, the weaving was completed using the pattern, with borders and long fringes added as the final touch.

Chilkat robes played a key part in the Tlingit trading economy, going for as much as $30 in the mid-1800s, an impressive price for that time. In the 20th century, however, the art of making the Chilkat robe has all but died out, with only a few committed weavers remaining.

PRESERVING IT FOR THE FUTURE

Preserved as a result of a fortunate series of circumstances, the totem poles at Sitka National Historical Park provide a sort of outside museum where visitors can view more than a dozen examples of the boldly carved cedar poles—one of the most magnificent and impressive art forms of the Northwest Coast Native Americans. Given to Governor John Brady of Alaska by Tlingit and Haida villagers of Prince of Wales Island for display in the 1904 Louisiana Purchase Exposition held in St. Louis, Missouri, these works of art traveled the long distance to represent Alaska at that centennial event. The following year they were included as part of Alaska's Lewis and Clark Exposition in Portland, Oregon. Fulfilling his promise to respect, protect, and preserve these poles, Brady then brought them to Sitka, where they were placed along what was then known as the old Russian walk to Indian River, where they remain today, carefully maintained and conserved as part of the national park. Some are copies of originals, which have been placed in storage to halt deterioration. Although the original totem poles didn't come from Sitka, they are very much part of the Tlingit tradition of this region.

Sitka National Historical Park is Alaska's oldest federally designated park, established in 1910 to commemorate the Battle of Sitka. An additional unit of the park preserves the Russian bishop's house in the town of Sitka. It is one of the few surviving examples of Russian colonial architecture in North America and dates from the 1840s.

EXPLORING ♦ FURTHER

Books about the Tlingit and Northwest Indians

Ackerman, Maria. *Tlingit Stories*, with story contributions from Austin Hammond, Sr., et al. Anchorage, Alaska: AMU Press, 1975

Emmons, George Thornton. *The Tlingit Indians*, edited with additions by Frederica de Laguna and a biography by Jean Low. Seattle, Wash.: University of Washington Press; New York: American Museum of Natural History, 1991.

Halpin, Marjorie M. *Totem Poles: An Illustrated Guide.* Vancouver: University of British Columbia Press, 1981.

Holm, Bill. *Northwest Coast Indian Art: An Analysis of Form.* Seattle, Wash.: University of Washington Press, 1965.

Liptak, Karen. *Indians of the Pacific Northwest.* The First Americans Series. New York: Facts On File, 1991.

Osinski, Alice. *The Tlingit.* New True Books. Chicago: Children's Press, 1990.

Paul, Frances Lackey. *Kahtahah: A Tlingit Girl.* Illustrated by Rie Munoz. Seattle, Wash.: Northwest Books, 1996.

Related Place

Fort Ross State Historic Park
19005 Coast Highway 1
Jenner, CA 95450
(707) 847-3286

Unaware that their hunting and gathering lifestyle would be changed forever, a group of Kashaya Pomo Native Americans assembled to watch the Russian ship that landed on the coast of California in March 1812. These Russians had come to hunt sea otter, to grow wheat and other crops for the Russian settlements in Alaska, and to trade with Spanish California. Alaskan Kodiak Island hunters in kayaks, who had come south with the Russians, did much of the otter hunting. On a scenic bluff they founded Fort Ross (probably short for *Rossiya,* or "Russia"), located 11 miles northwest of what is now the town of Jenner on Highway 1, about a two-hour drive from San Francisco. Many of the structures remain or have been reconstructed much as they were when the Russians lived there.

Little Bighorn Battlefield National Monument

FLEETING TRIUMPH OVER THE UNITED STATES GOVERNMENT
Crow Agency, Montana

Date: June 25–26, 1876

Site of the Battle of the Little Bighorn, where Sitting Bull, Crazy Horse, and others led Lakota and Northern Cheyenne warriors against the Seventh Cavalry commanded by George Armstrong Custer.

Address:
Little Bighorn Battlefield National Monument
P. O. Box 39
Crow Agency, MT 59022
(406) 638-2622
Fax: (406) 638-2623

> *Of all the confrontations between American Indians and European Americans, the Battle of the Little Bighorn is one of the best known and most significant clashes of the two cultures. It was also one of the bloodiest.*

On a scorching June Sunday in 1876, hundreds of Indian warriors converged on a grassy ridge rising above the valley of Montana's Little Bighorn River. On the ridge five companies of United States cavalry, about 225 officers and troopers, fought desperately but hopelessley against many times their number. When the guns fell silent and the smoke and dust of battle lifted, no soldier survived.
—Robert M. Utley, National Park Service

♦ ♦ ♦ ♦ ♦

Photo of Custer Battlefield, taken ca. 1883   (Courtesy of the National Park Service)

THE AMERICAN INDIAN EXPERIENCE

Originally erected to commemorate United States soldiers who died there, the Custer Battlefield National Monument is now called the Little Bighorn Battlefield National Monument. Nevertheless, the white marble tombstones still stand against the waving grasses on the knoll where forces led by Lieutenant Colonel George A. Custer fell to their deaths at the hands of angered Lakota* and Northern Cheyenne warriors led by Sitting Bull, Crazy Horse, and Gall.

For Custer, once the dashing, flamboyant young military hero who enchanted much of the nation, it was his last battle. For the Indians, it was an unmitigated triumph in the struggle against white encroachments and injustices. For the descendants of combatants on each side of this historic battle, this place holds great poignancy. Most of the U.S. Army soldiers who died here are buried in a common grave on Last Stand Hill. The Indians who fell were removed by their families.

In years prior to the battle at Little Bighorn many conflicts had occurred between the U.S. military and Lakota warriors, Sitting Bull in particular. The full context of the situation, however, was much longer in the making. Originally, the Teton Lakota—one of four branches of Lakota—had lived around the headwaters of the Mississippi. Made up of seven separate bands— the Hunkpapa, Oohenonpa, Oglala, Brulé (Sicangu), Sihasapa, Itazipco (Sans Arc), and Miniconjou—their people had been pressured out of that area by Chippewa (Anishinabe) armed with muskets provided by white traders. In the late 18th century the Teton had retreated to the Missouri River and even farther westward. They, in turn, had overrun the weaker tribes in these areas, and the Teton, with their swift horses, swept across the Great Plains.

The Cheyenne had experienced similar pressures, splitting into two groups, the Northern and Southern Cheyenne. The Northern Cheyenne migrated farther to the southwest than the Lakota, occupying the upper Platte River region.

But there was one major drawback to this arrangement. The land the Lakota and the Northern Cheyenne had moved onto was Crow territory. And relations between the Lakota in particular and the Crow were characterized by long-term hostility, as the Lakota helped themselves to buffalo that had once fed the children of the Crow.

*Historically the Lakota were known as the Sioux, a name taken from an Ojibwa word meaning "enemies."

Then came another element—this time a wave of white settlers. Whereas the Lakota had been taking over territory, now they were forfeiting it, and this time they had nowhere to go. Finally, in 1868 the Lakota signed a treaty, which agreed to an "unceded territory," one on which the Lakota could roam free, unhampered by the presence of whites. But from the beginning the government was unhappy with the arrangement. Then gold was discovered in the Black Hills, not on the unceded territory, but on the Great Sioux Reservation. By the spring of 1875, goldseekers were stampeding to the area. The government tried to buy the Black Hills from the Lakota, who considered it a sacred place. But the young warriors who spent their summers with Sitting Bull would have none of it. The Commissioner of Indian Affairs issued an ultimatum: Come in to the agency by January 31, 1876 or be branded hostile and face being pursued by the army. The Indians ignored the demand, and the army prepared to pursue, completely miscalculating that numbers ranged against them might be an important consideration in their plans.

The Indians gathered forces. During a Sun Dance in May, Sitting Bull had a prophetic vision: He saw many dead soldiers "falling right into our camp."

The army, meanwhile, had been divided by circumstances. The campaign strategy had called for three separate prongs, but one of the forces, under General George Crook, had been knocked out of the picture when they encountered a large Cheyenne-Lakota force on the Rosebud River and were forced to withdraw. That left the two others—one under Colonel John Gibbon and the second under General Alfred H. Terry. The plan was to converge on the main body of Native Americans gathered in southeastern Montana under the leadership of Sitting Bull, Crazy Horse, and others. This camp included Indians from a number of groups, also including Blackfeet and Hunkpapa. Terry sent Custer, who was in his command, off in one direction with the Seventh Cavalry to approach an area known to the Indians as Greasy Grass along the Little Bighorn River from the south. Terry and Gibbon would loop around and come in from the north.

Custer had about 600 men with him when the Seventh Cavalry located the Indian camp at dawn on June 25. No one knows quite why Custer proceeded as he did, but he must have underestimated the size and fighting expertise of the Lakota and Cheyenne forces. There may have been as many as 7,000 Indians in the camp, including about 1,800 warriors. Custer divided his regiment into four battalions, one of which stayed behind to protect the

Reno Battlefield 10 years after the Battle of the Little Bighorn (Courtesy of the National Park Service)

pack train. He retained five companies under his immediate command and assigned three companies each to Major Marcus A. Reno and Captain Frederick W. Benteen. Custer and Reno headed toward the Indian village while Benteen was to scout the bluffs. Custer and Reno took opposite ends of the encampment, but Reno was forced into an ungraceful retreat by a large band of warriors, then went looking for Custer, who from the sound of the gunfire appeared to need help. But he couldn't find Custer's command, and he and Benteen wound up under siege. Reno and Benteen were able to hold their position until Gibbon and Terry arrived with their columns.

No one is certain of Custer's movements that day, but Northern Cheyenne chief Two Moons recalled that "the shooting was quick, quick. Pop-pop-pop very fast. Some of the soldiers were down on their knees, some standing.... The smoke was like a great cloud, and everywhere the Lakota went the dust rose like smoke. We circled all around him—swirling like water around a stone. We shoot, we ride fast, we shoot again. Soldiers drop, and horses fall on them."

Reno's Retreat, *drawn by* White Bird, *a Northern Cheyenne warrior who participated in the* Battle of the Little Bighorn (United States Military Academy, West Point. Courtesy of Little Bighorn Battlefield National Monument)

Standing today on the windswept hills above the Little Bighorn, the strange mixture of exultation on one side and death and defeat on the other seems to fill the air with a somber smoke like gunpowder.

A CLOSE UP — WHERE ARE OUR LANDS?

The joy of victory was short lived for the Indian forces. The United States government—outraged at the obliteration of the entire command of a national military hero—retaliated swiftly with both political and military pressure. At the reservations, military rule was enforced, and Indian men were deprived of their mounts and disarmed—even though many of them could prove they had not left the reservation. In September 1876, the chiefs responded to demands by the U.S. Congress and signed away the gold-laden Black Hills in Dakota Territory (now South Dakota).

From Little Bighorn, groups of Native Americans had fled in separate directions, and in the following months, the U.S. Cavalry sought to press in on them, one by one. On November 25, 1876, some 1,100 men under the command of Ranald S. Mackenzie attacked the village of Cheyenne chief Dull Knife, who had been joined by Little Wolf on the Red Fork of Powder River. Divested of supplies, food, tipis, and ponies, those who survived were broken in spirit, even though Crazy Horse and his band took them in. By spring, even Crazy Horse saw that the end had come, leading 900 to 1,100 Lakota into the Red Cloud agency in Nebraska in a solemn, 2-mile processional. There, he threw his weapons on the ground. Within six months, he was dead, killed by a bayonet in a guardroom struggle.

Sitting Bull fled north to Canada from Little Bighorn with his band of 400 Hunkpapa. But foraging for food proved difficult north of the border, the Canadian government was less than welcoming, and the border was guarded so carefully that they could not hunt buffalo on the wide-open plains of Montana or meet with other Lakota.

Still, Sitting Bull resisted efforts made by U.S. government officials to lure him back, from time to time making his position clear with announcements like the following:

> When I was a boy the [Lakota] owned the world; the sun rose and set on their land.... Where are the warriors today? Who slew them? Where are our lands? Who owns them?... What law have I broken? Is it wrong for me to love my own?

But gradually defeated, factions of his band drifted into the reservations little by little, until finally, five years after Little Bighorn, Sitting Bull gave in. He surrendered on July 19, 1881, at Fort Buford, North Dakota, with his 187 remaining followers.

"I wish it to be remembered," he declared, "that I was the last man of my tribe to surrender my rifle, and this day have given it to you."

PRESERVING IT FOR THE FUTURE

Little Bighorn Battlefield National Monument lies within the Crow Indian Reservation in southeastern Montana. The area was established as a National Cemetery January 29, 1879, three years after the battle, and it was declared a National Monument March 22, 1946. Four audio stations provide historical information, and during the summer months, rangers give daily talks about the battle, the life of the frontier soldier, and the culture of the Plains Indians. Exhibits in the visitor center also display facts about the battle and culture of the Plains Indians. Visitors can tour the battlefields by car, by bus, or on foot.

EXPLORING ◆ FURTHER

Books about the Battle of the Little Bighorn and Related Topics

Connell, Evan S. *Son of the Morning Star: Custer and the Little Bighorn.* San Francisco: North Point Press, 1984.

Greene, Carol. *Black Elk: A Man with a Vision.* Rookie Biographies. Chicago: Children's Press, 1990.

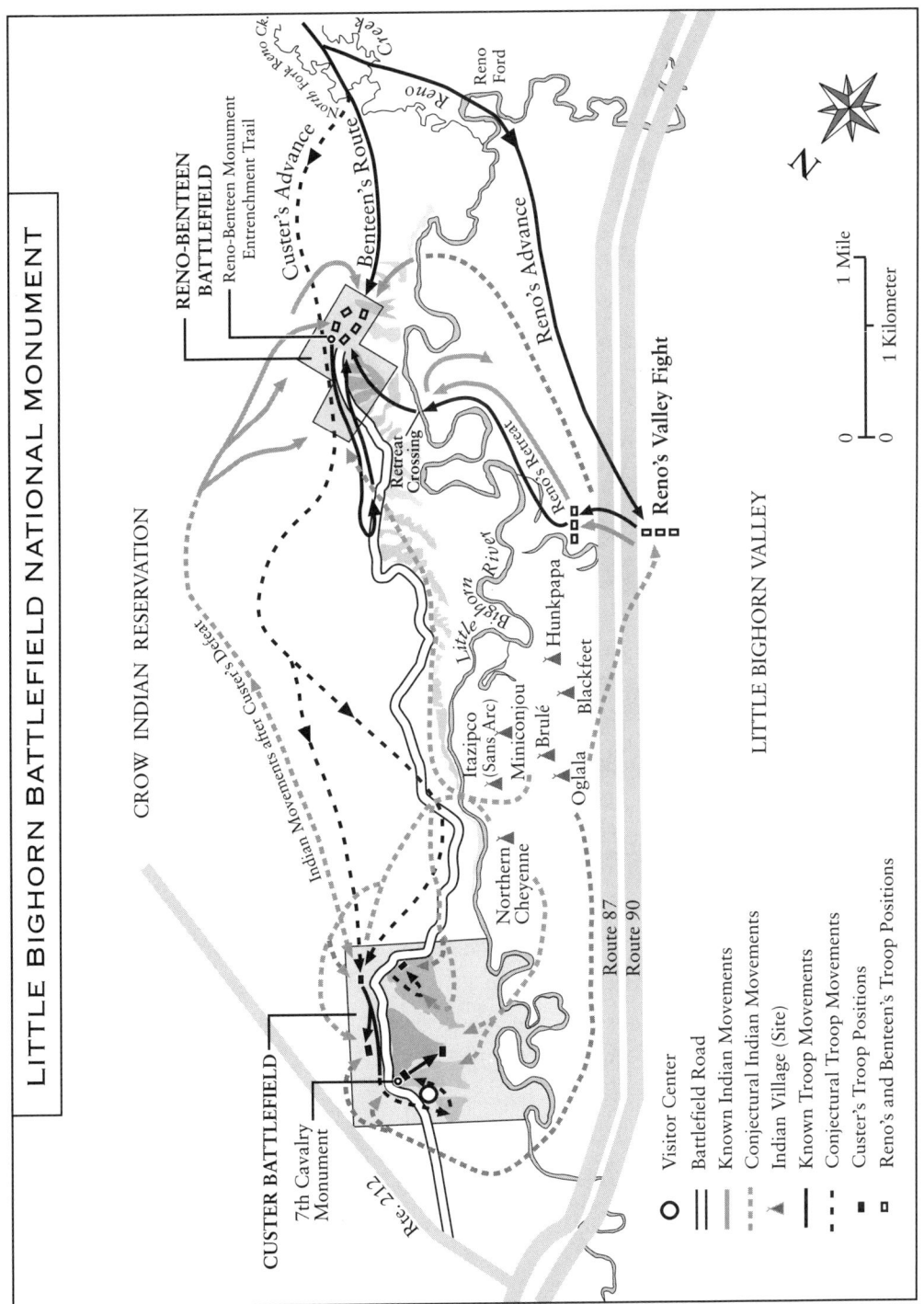

Hoig, Stan. *The Cheyenne*. Indians of North America Series. New York: Chelsea House, 1989.

Stein, R. Conrad. *The Battle of the Little Bighorn*. Cornerstones of Freedom. Chicago: Children's Press, 1997.

Taylor, William O. *With Custer at the Little Bighorn*. New York: Viking-Penguin, 1996.

Related Place

Crazy Horse Memorial
Avenue of the Chiefs
Crazy Horse, SD 57730-9506
(605) 673-4681
Fax: (605) 673-2185
E-mail: Memorial@CrazyHorse.Org

In 1947, sculptor Korczak Ziolkowski began work on this giant monument to the spirit of Crazy Horse, who fought at Little Bighorn and valiantly resisted the loss of Indian lands and way of life. For many, Crazy Horse represents pride and valor, which Ziolkowski has attempted to capture in his sculpture—a carving in the round of the revered chief on horseback. The giant undertaking, when finished, will replace most of the peak of Thunderhead Mountain in the Black Hills near Custer, South Dakota. It is predicted to be the largest sculpture in the world, measuring 563 feet tall and 641 feet long. Ziolkowski spent the remainder of his life on this project, in which he involved his entire family, and his wife, Ruth, took over its direction after his death in 1982. Visitors can see Ziolkowski's reduced-scale model, his studio, home, and workshop, and the emerging sculpture in progress. The Indian Museum of North America at the site features a vast collection of Native American art, artifacts, and crafts.

Nez Perce National Historical Park

HOMELAND IN THE NORTHWEST
Spalding, Idaho

AT A GLANCE

Dated from: ca. 1700–1877

Park relating to the Nez Perce, with sites in four states: Idaho, Montana, Oregon, and Washington

Part of the Columbia Cascades Cluster, this park, with headquarters in Spalding, Idaho, encompasses four sites administered by the National Park Service, twenty more by cooperative agreements, and another fourteen related sites authorized in Montana, Oregon, and Washington.

Big Hole National Battlefield, in Wisdom, Montana, is a memorial to the people who fought and died there on August 9 and 10, 1877. It is part of the Nez Perce complex but maintains separate headquarters.

Address:

Nez Perce National Historical Park
P.O. Box 93
Spalding, ID 83551
(208) 843-2261

Big Hole National Battlefield
P.O. Box 237
Wisdom, MT 59761
(406) 689-3155

> *This group of sites together celebrate the culture of the Nez Perce people, commemorate their history, and examine their relationship to North American history.*

Hear me, my chiefs, I am tired; my heart is sick and sad.
From where the sun now stands I will fight no more forever.
—Chief Joseph of the Nez Perce, 1877

♦♦♦♦♦

On the high, windswept desert plateaus of eastern Oregon, Washington, central Idaho, and western Montana is the land where the Nez Perce (who now call themselves Nimiipu) have lived for more than 11,000 years. Dubbed the Nez Percé (French for "pierced nose"—pronounced

The nomadic Nez Perce (Nimiipu) once freely roamed the vast high plateaus and valleys of this territory. (Courtesy of the National Park Service)

According to Nez Perce legend, Coyote hurled his fishnet high onto the banks on this side of the river (dark horizontal area in the middle of the picture). (Courtesy of the National Park Service)

[NAY pair SAY]) by French-Canadian trappers, they became known as the Nez Perce, pronounced [NEZ PURS]. In this, their traditional homeland, they interacted with the environment and landscape, developing a life that harmonized with their surroundings and a tradition of courage and spiritual strength of which they were proud. Once, a time long ago, grandparents told the Nez Perce children, the world was inhabited only by animals. In a creation story that was common throughout the Northwest, they told of animals that spoke like humans and had human characteristics. By one of the waterways lived a fierce and terrible monster who devoured incautious animals and kept them all in fear. But the bold and courageous Coyote—hero of the culture—jumped down the monster's throat and sawed up his heart with a flint. When the monster was dead, he cut up its body into many parts, and of each part he created a tribe. The Nez Perce, the children were told, sprang from the very heart of the monster, making those who came from this lineage the boldest and most courageous of all. Today visitors can see the site near

Kamiah, Idaho, where the Nez Perce say the remains of the "Heart of the Monster" and "Liver of the Monster" have been turned into a geologic formation of basaltic rock.

Nez Perce tradition abounds with tales of the exploits of Coyote. In another story, he once was fishing with a large net in the Clearwater River. Black Bear happened to come by and his actions angered Coyote, who mightily stalked out of the water and hurled his fishnet high up on the hill to the south. Then he hoisted up Black Bear and flung him far up the opposite hill, almost to its brow, on the other side of the river. As final punishment, he turned Black Bear into black stone, and you can see both Coyote's Fishnet and Black Bear still there today, about 7 miles east of Lewiston, Idaho, near the Clearwater River.

Centered in the Nez Perce homeland, the Nez Perce National Historical Park encompasses 38 separate park sites in four states—Oregon, Washington, Montana, and Idaho—reflecting in this way the expansiveness of the

It is said that in a snit, Coyote hurled Black Bear high up onto the brow of the hill overlooking the river on the other side, where you can still see him today, the dark area just below the peak. (Courtesy of the National Park Service)

Nez Perce territory. From this vantage point the Nez Perce witnessed the march of history and change that transformed their lives. It is a different viewpoint from the one gained, for example, standing at the Gateway Arch in St. Louis.

Recent Nez Perce history began in about 1700, with the introduction of the horse, after which the people called Nez Perce became noted for the large horse herds they raised. Their culture transformed from the original Plateau culture to a more nomadic way of life having more Plains traits, as they became renowned as breeders of the Appaloosa horse, known for its spunk and stamina. Their horses were so important, in fact, that the Nez Perce gauged the wealth of a family by the number of horses it owned. Groups of villages were led as a band by a council of headmen and a war leader.

They lived in small, scattered settlements of 10 to 75 related families, their villages composed of several different types of homes, including excavated pit houses, tipis, and longhouses, which were mat-covered sheds sometimes 100 to 150 feet long. The tipi in particular was popular, as it was throughout North America, because it was easy to transport. To move a tipi, occupants simply disassembled it and lashed two of the supporting poles to either side of a horse, letting the ends drag on the ground, to form a roughly triangular frame, called a travois. On the travois, they lashed mats, hides, and other family possessions. Once arrived at the new campsite, they could bind together the long lodge poles near their tops. Then they stood the poles up and spread them outward to form the outline of a cone. The builders leaned other poles against this framework to strengthen it, and then tied layers of tule (bulrush) mats to the poles like shingles. Where the elevation was high and the nights were cold, they also added a layer of skins sewn together to break the wind. Some Plains tribes used buffalo hides instead of mats.

In winter, for warmth, several families often lived together in the longhouses. Fires were arranged in rows down the middle of the house, about 10 or 12 feet apart, with two families to each fire. Personal belongings and bedding were arranged along the walls. The Nez Perce elevated the beds, using layers of dry grass and the inner bark from cottonwood trees to form a soft mattress.

By the early 1800s, the Nez Perce prophets began to caution about the arrival of strange people from the east who would bring with them many changes, as William Clark and Meriwether Lewis made their famous exploratory trek along the Lolo Pass through the Great Divide and into Idaho in 1805. By 1811, the advent of the fur trade into Nez Perce territory began to change their way of life. Presbyterian missionaries followed in 1836, and more and more settlers flowed into areas that had always been available to the Nez Perce for the hunting and fishing on which they depended for their food. The two cultures were incompatible—the settlers, having no perception of Indian values, thought of the Indians as godless savages, while the Nez Perce felt that the Euro-Americans were like greedy, undisciplined children and they didn't understand the white settlers' need for structure and control.

By 1855, the white settlers, who believed possession of this land was their "Manifest Destiny," had pressured the government in Washington to force land cessions from the Indians. Desiring peace, the tribe agreed to a treaty that confined them to a spacious reservation that included much of their ancestral land. A provision of the treaty protected them, stipulating that non-Indians could live on the reservation only with the Nez Perce's consent.

Then in 1860, gold was discovered on the Nez Perce reservation. Suddenly settlers and miners were hungry for huge pieces of the Nez Perce's land, and they pressed for a new treaty in 1863 that would reduce the reservation to one-tenth its original size, including only the lower portion of the Clearwater and the grasslands southwest of this 100-mile section of the river. The Nez Perce from the Salmon River region, Hell's Canyon on the Snake River, the Wallowa Lake area, and along the Snake would all lose their lands. A council to discuss the terms of the treaty was called in the spring of that year by officials from Oregon State and Washington Territory. It was attended by representative bands from the Nez Perce nation, some of whom had little to lose and began to give in. Others voiced strong resistance to the proposed extreme reduction of their lands. In a late-night debate among the Nez Perce, a band led by a chief named Big Thunder emotionally withdrew from the discussion, declaring the Nez Perce Nation dissolved. Captain George Currey, who witnessed this meeting, later wrote, "I withdrew my detachment, having accomplished nothing but witnessing the extinguishment of the last council fires of the most powerful Indian nation on the sunset side of the Rocky Mountains." The document that was signed took 6.9 million acres from the Nez Perce, and the federal government acted on it, claiming that the majority of the tribe was represented by the few who signed (nearly all of whom already lived in the new reservation area).

Ultimately five groups who refused to sign and to abide by the new treaty—about a third of the tribe—became known as the non-treaty Nez Perce. Tuekakas, father of Joseph of the Wallowa Nez Perce, was among those who refused to sign. As remembered by his son, he said to the government representative, "I have no other home than this. I will not give it up to any man. My people would have no home. Take away your paper. I will not touch it with my hand." Not long thereafter, Joseph became chief of his band in the place of his father, who had become blind and ill.

The non-treaty bands remained in their homelands for several years. But finally, in 1877, the government began—in response to increasing pressure from settlers—to insist that the non-treaty Nez Perce move to the reservation. General Oliver O. Howard received instructions to make sure the order was

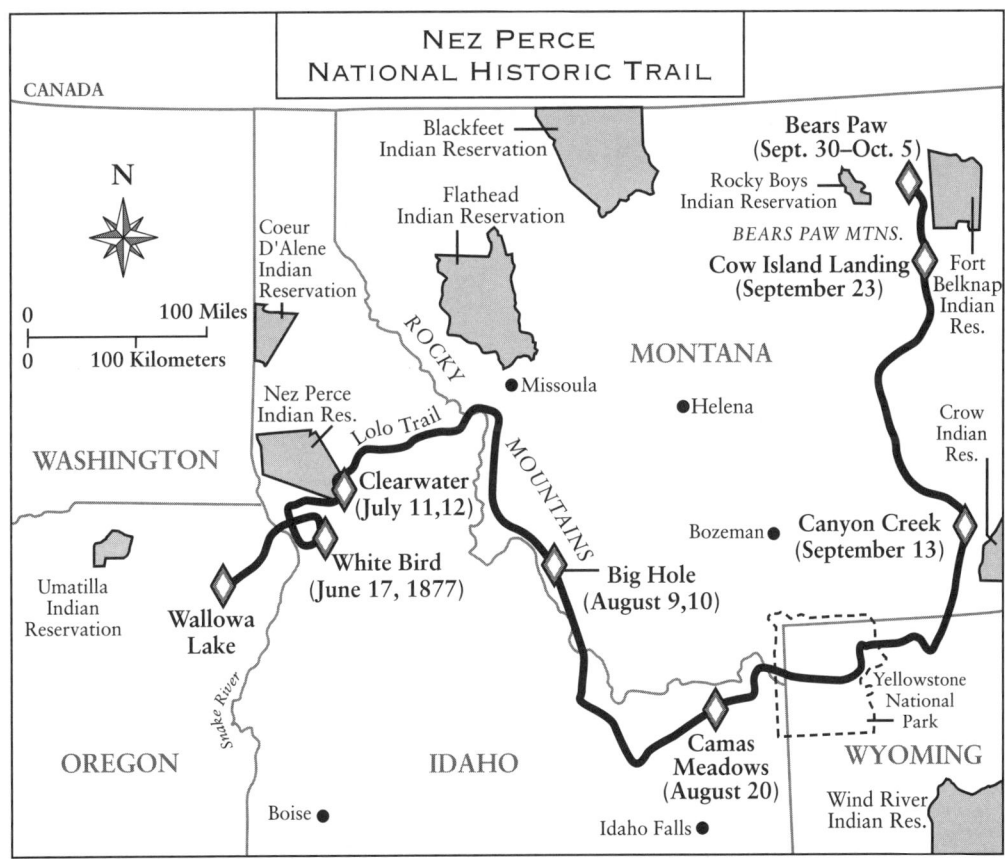

The 1,600-mile journey taken by non-treaty Nez Perce as they fled U.S. soldiers in 1877

NEZ PERCE NATIONAL HISTORICAL PARK

obeyed. In mid-May, Howard issued an ultimatum: The non-treaty Nez Perce must be on the reservation within 30 days.

Chief Joseph objected. Speaking for the rest of the bands, he said, "Our stock is scattered and Snake River is very high. Let us wait until fall, then the river will be low." Refusing this appeal, General Howard threatened to use force if the Nez Perce did not meet the deadline. So began an incredible journey that would take Joseph and his people halfway across the continent. Reluctantly, the non-treaty chiefs persuaded their people to obey the orders. They rounded up what they could of their livestock in the short time they had, packed up as much as they could and struggled across the Snake and Salmon rivers, which were swollen with spring rains. They almost made it to the reservation in time, when three of their young warriors began a revenge attack on white settlers who had killed members of their families. Before the confrontation was over, other Nez Perce had joined them, and they had killed seventeen settlers in two days of raids.

Most of the non-treaty Nez Perce saw themselves as vulnerable to retaliation and retreated to White Bird Canyon, where they knew the terrain would help them defend against a surprise attack. General Howard sent out 99 cavalry and 11 civilian volunteers to put down the uprising on June 17, but the small band of poorly armed Nez Perce repulsed the attack and the U.S. troops retreated, suffering heavy losses. Now the non-treaty Nez Perce were on the run. The following month involved numerous skirmishes and encounters with soldiers. General Howard brought in more troops from up and down the West Coast to encircle the elusive Nez Perce. On July 11 another battle took place on the Clearwater River, but the Nez Perce continued on across the Lolo Pass into the Bitterroot Valley in Montana, breathing relief that they had made it beyond Howard's easy reach.

The Nez Perce arrived in the lush Big Hole Valley on the morning of August 7. Their principal leader, Chief Looking Glass, stopped at an old campsite, where the band set up their tipis. But they didn't know that a second force, composed of 162 men from five Montana forts, had been ordered to the scene under the command of Colonel John Gibbon. Gibbon's scouts spotted the tipis on the afternoon of August 8, and Gibbon marshaled his attack the following day, although an accidental discovery of troops by a lone Nez Perce forewarned the Indians. The battle was bloody, with women and children

Nez Perce campsite at the Big Hole National Battlefield, Montana (Courtesy of Jock Whitworth/National Park Service)

shot indiscriminately as they awakened from their sleep. But Nez Perce snipers forced the soldiers to retreat with many casualties.

During the attack, some of the soldiers had just managed to haul a mountain howitzer into position in the dense pine forest and had fired off only two rounds when Nez Perce warriors swooped in, captured the gun, dismantled it and sent the wheels rolling down the hill. A replica of the howitzer stands today at the site of capture.

The Nez Perce hurriedly buried their dead and headed south, leaving many of their belongings behind, including the lodge poles of their tipis, which you can still see at their Big Hole encampment site. Final shots were fired on August 10 as the last of the Nez Perce warriors departed to join their people. The battle was over, and technically won by the Nez Perce—the military had losses of 29 dead and 40 wounded—but 60 to 90 members of the tribe were dead, many of them women and children and old people. They realized they were fleeing for their lives and eventually they decided to head toward

Canada to join Lakota chief Sitting Bull, who had gone there after the Battle of the Little Bighorn the previous year.

Most never got there, though. They headed south to Shoshone country, hoping to pick up warriors to replace those they had lost, meeting further confrontations at Birch Creek and Camas Meadows, in Idaho. They crossed the Yellowstone in Wyoming and fought two more major battles in Montana before finally, on September 30, they were surprised by Colonel Nelson A. Miles near the Bear Paw Mountains of Montana. After five days of fighting, interspersed with negotiations, and the deaths of four chiefs, Chief Joseph surrendered to Miles. They had traveled nearly 1,300 miles. Eight hundred non-treaty Nez Perce had begun the journey. Only 431 remained in Big Paw to surrender. But not all of the rest were killed—some 200 succeeded in reaching Canada, while some had hidden in the hills. But the Nez Perce were tired. Many fighting men were gone. And the losses of women, children, and elders at Big Hole had taken a heavy psychological toll.

To General Howard's appeal for an end to fighting, Chief Joseph sent this message:

> . . . I am tired of fighting. Our chiefs are killed. Looking Glass is dead. Toolhoolhoolzote is dead. The old men are all dead. . . . It is cold and we have no blankets. The little children are freezing to death. My people, some of them, have run away to the hills and have no blankets, no food; no one knows where they are—maybe freezing to death. I want time to look for my children and see how many of them I can find. Maybe I shall find them among the dead. Hear me my chiefs. I am tired. My heart is sick and sad. From where the sun now stands I will fight no more forever.

For Joseph and many of his people, however, the journey was not over. New orders came to send them across North Dakota, northern Minnesota, Nebraska, and Iowa to Fort Leavenworth in Kansas for confinement, traveling by railroad. In Kansas Joseph embarked on a long and frustrating campaign to secure the return of his people to Idaho, even traveling to Washington, D.C. to make his case in person. While there he spoke out clearly about the racism and injustice he had seen. "If the white man wants to live in peace with the Indian he can live in peace," he said. "There need be no trouble. Treat all men alike. Give them all the same law. Give them all an even chance to live and grow. All men were made by the same Creator. They are all brothers."

Finally, after eight years of lobbying, in 1885, 118 Nez Perce were returned to the reservation in Idaho. The remaining 150, including Joseph, were sent to the Colville Reservation in the state of Washington.

A CLOSE UP: LOSING THE BATTLE FOR A WAY OF LIFE

The final event in the Indian struggle to retain the lands and life that had once been theirs came in the winter of 1890–91.

The stage was set, completely without intention, by a Northern Paiute holy man named Wovoka, who taught that a brighter future awaited Indians and that by performing a ritualistic dance called the "Ghost Dance," they could help bring it about. Wovoka's wisdom came to him in a vision that occurred during an eclipse of the sun. "When the Sun died," he said, "I went up to heaven and saw God and all the people who had died a long time ago. God told me to come back and tell my people they must be good and love one another, and not fight or steal, or lie. He gave me this dance to give my people." As the mythology evolved and spread to other tribes, those who practiced the Ghost Dance religion said there would be a new era in the future, when all those who had died would return and there would be no white people. Shirts worn during the dance, called Ghost Shirts, would also make the wearer impervious to enemy bullets.

The new religion gave hope to many Indians, who traveled as delegates from all over the country to Wovoka's conferences, returning with renewed inspiration and fervor which they passed on to practitioners in the reservations.

U.S. government officials, meanwhile, were unsure of the meaning of these comings and goings and the fevered dancing, but they began to become concerned about the Ghost Dance movement and began building up military installations near reservations, in case trouble erupted.

During this time, a group of the Lakota people, who lived in South Dakota, started to dance the Ghost Dance. On the Cheyenne River Indian Reservation, the government ordered Chief Big Foot's arrest for his support of the dance. Although Big Foot and his band of more than 200 people evaded capture at first, the army succeeded in rounding them up on December 28, 1890 and escorted them to the military camp near Wounded Knee Creek. The next morning the cavalry commander, Colonel James W. Forsyth, called a council to collect the band's weapons. Although surrounded by cavalry and infantry backed up by Hotchkiss cannons, the Lakota gave them up with great reluctance. A few young warriors retained some weapons beneath their blankets as they stood in the council circle. When pressed, though, most of these men gave them up, except for one, who brandished his in the air. A shot went off, and a tremendous barrage of army rifle and artillery fire ensued. Many were killed instantly. Others were mowed down like grass as they tried to escape to a nearby ravine. In the end, more than 200 Lakota men, women, and children lay dead. Thirty soldiers also lost their lives—mostly due to cross-fire.

Later, a burial party returned to the snow-covered terrain to find and bury the dead. According to the account Dr. Charles A. Eastman, an Agency physician (of Santee Sioux heritage) who accompanied the party: "Fully three miles from the scene of the massacre we found the body of a woman completely covered with a blanket of snow, and from this point we found them scattered along as they had been relentlessly hunted down and slaughtered while fleeing for their lives. . . . When we reached the spot where the Indian camp had stood, among the fragments of burned tents and other belongings we saw the frozen bodies lying close together or piled one upon another."

With those tragically killed that day, the promise of the Ghost Dance also died for the Lakota people, and they were forced to accept reservation life. For the United States, it was the end of the "Indian Wars." But the Lakota say that the sacred hoop of their nation—which they equate with the life and health of their nation—broke at Wounded Knee.

PRESERVING IT FOR THE FUTURE

Nez Perce National Historical Park was created on May 15, 1965 and continues to undergo development, as new interpretative programs and signage are added. The main park headquarters are located in Spalding, Idaho, although secondary headquarters are coordinated from Big Hole, Montana, where the battle of 1877 took place and which is designated as a national park in its own right.

EXPLORING ♦ FURTHER

Books about the Nez Perce

Brown, Mark H. *The Flight of the Nez Perce.* 1967. Reprint. Lincoln, Neb.: University of Nebraska Press, 1982.

Hines, Donald M. *Tales of the Nez Perce.* Fairfield, Wash.: Ye Galleon Press, 1984.

Sanford, William R. *Chief Joseph: Nez Perce Warrior.* Native American Leaders of the Wild West. Hillside, N.J.: Enslow Publishers, 1994.

Scott, Robert A. *Chief Joseph and the Nez Percés.* Makers of America Series. New York: Facts On File, 1993.

Taylor, Marian W., et al. *Chief Joseph: Nez Perce Leader.* North American Indians of Achievement Series. N.Y.: Chelsea House, 1993.

Yates, Diana. *Chief Joseph: Thunder Rolling Down the Mountains.* Unsung Americans Series. Staten Island, N.Y.: Ward Hill Press, 1992.

Related Place

Badlands National Park
(for information)
Journey to Wounded Knee
P.O. Box 6
Interior, SD 57750

(605) 433-5361
Wounded Knee Monument
Pine Ridge Reservation
Pine Ridge, SD 57770

Today only a sign and mass grave mark the site of the Wounded Knee Massacre on South Dakota's Pine Ridge Indian Reservation, the location of one of the most important and tragic events in American Indian–United States relations. But visitors to Wounded Knee can view the creek, grass-covered hills, and ravine where the massacre occurred as they contemplate its meaning. They can also sense the sacredness of the site to Lakota (Sioux) people, many of whom view the entire area as a burial ground. Nearby Badlands National Park can supply directions and a map that helps visitors visualize the council circle, Indian camp and surrounding troops, cannon, and cavalry just prior to the massacre that took place here December 29, 1890.

Across the road from the historical marker is the site of the church occupied in 1973 by members of the American Indian Movement (AIM) to protest broken treaties and wrongs against Indians. The occupation resulted in the death of two Indians. The church cemetery still exists, although the church itself burned down.

Alcatraz Island

TAKING A STAND IN THE TWENTIETH CENTURY
San Francisco Bay, California

AT A GLANCE

Built: 1854
Seized by Indian activists: November 22, 1969

Site of American Indian protest, 1969–71

A rocky island, 1,650 feet long and 450 feet wide, located in the San Francisco Bay in western California. Used as a prison from 1854 to 1963, in 1969 it became the unlikely site of an Indian protest occupation.

Address:
Alcatraz Island
Golden Gate National Recreation Area

Headquarters:
Fort Mason, Building 201
San Francisco, CA
(415) 705-1042/(415) 556-0560

> *The Rock had always been a prison, until 1963, when the federal government gave it up as too expensive to run. Six years later, it became the focal point of a dramatic gesture intended to make a point to the American public and to fire the imaginations of American Indians of every tribal background.*

Seizing the Rock had been a stunning turning point in the history of Indian protest.

—Paul Chaat Smith and Robert Allen Warrior,
in *Like a Hurricane*, 1996

♦ ♦ ♦ ♦ ♦

As boats filled with young Indian activists approached the rocky dungeon island in the middle of San Francisco Bay, they received the order to turn around. Compliantly, they turned to head back to the bayside docks, when one lone Mohawk suddenly leaped into the icy waters. Richard Oakes swam for his life—the currents were strong, the water choppy. And in that moment he realized it could be done. If he could gather together enough students and activists, and if they moved cunningly, they really could occupy the Rock, the formidable prison island that the federal government had abandoned since 1963.

It was November 9, 1969. The occupation was short-lived that night, but Oakes, then a student at San Francisco State University, went to the American Indian Studies Center at UCLA where he recruited Indian students. Because many tribes were represented, they called themselves "Indians of All Tribes," and on November 22, a group of approximately 100 Indian people occupied Alcatraz. Eighty of those were Indian students from UCLA. Citing a treaty of 1869, they claimed the tiny 1,650-foot-by-450-foot island for the Indians of All Tribes. The government initially insisted that the Indians leave the island, but negotiations ensued, and the occupation gained considerable favor in the media.

An Indian spokesperson presented Native American plans for the island at a press conference on Alcatraz. (By permission, San Francisco Chronicle. From the Photo Collections of Golden Gate National Recreation Area National Park Service No. GoGA-2316, Interpretative Photograph Collection)

Not all Native Americans thought the venture made any sense, however. The occupiers contended that they wanted to make the island into an Indian cultural center. They held news conferences, with maps and photographs, explaining their plans, but some Native Americans in Oklahoma said they wondered how the Indians of All Tribes thought they could maintain a facility on the island if the federal government had given up the place as impractical and too expensive to maintain.

Meanwhile, what had once been a formidable prison, incarcerating the most hardened of criminals, including Al Capone and Robert Stroud, the "Bird Man," became home to a hundred young men, women, and children. The Native American demands were firm. They wanted the deed to the island, an Indian university, a cultural center, and a museum.

Months passed, negotiations wavered, and leadership on the island foundered. Then tragedy struck when Oakes's 13-year-old step-daughter Yvonne fell down a stairwell to her death. After Yvonne's death, Oakes, who was already having difficulty leading the unwieldy group, left the island. Students had already begun to leave to return to classes in early 1970, but some stayed. Finally, with only fifteen occupiers remaining, the FBI removed the five women, four children, and six unarmed Indian men on June 11, 1971. The occupation was over, and its legacy was mixed.

But a new consciousness had been raised. While Alcatraz was not won, it proved a point, as artist "Indian Joe" Morris's poster title says. After the implementation of 1950s government policies of relocation and assimilation of reservation Indians into mainstream society, many Indians felt they were invisible—without a culture, without a language, without a land they could

As shown by this photo taken in the early 1960s, though isolated by strong currents and cold waters, the island's occupants always had the nearby San Francisco skyline squarely within view. (From the Photo Collections of Golden Gate National Recreation Area National Park Service)

call their own. As Two Moons, a Northern Cheyenne of a previous generation, once said, "Our old chiefs are fast dying away, and our old Indian customs soon will pass out of sight, and the coming generations will not know anything about us. . . ."

Alcatraz became a symbol of resistance and hope to the growing Indian, or Red Power, Movement, and it focused national attention on the problems and concerns of American Indians. It also set the stage for two more protest actions, the brief occupation of the Bureau of Indian Affairs in Washington, D.C. and an important demonstration at Wounded Knee in 1973, when 200 members of the American Indian Movement (AIM) occupied the site of Wounded Knee for 69 days, demanding a Senate investigation into the condition of Native Americans.

Directly or indirectly as a result of the occupation of Alcatraz, the official government policy of termination of Indian tribes (the end of legal recognition of tribes as sovereign bodies and thus the benefits and rights due them, in some cases including the removal from the reservations) was abandoned and replaced by a new official policy of Indian self-determination. During the Alcatraz occupation, President Richard Nixon agreed to support the return of Blue Lake and 48,000 acres of land to the Indians of the Taos Pueblo. Lands near Davis, California became the home of a Native American University, and several others have since been founded. And the occupation of the offices of the Bureau of Indian

Graffiti appeared on building walls soon after the occupation. Most evoked tribal pride, but some, like this one, conveyed deeply held anger about past conflicts with the U.S. government. (From the Photo Collections of Golden Gate National Recreation Area National Park Service No. GoGA-2316, Interpretative Photograph Collection)

ALCATRAZ ISLAND

Affairs led to hiring of Native Americans to work in this federal agency, which has so long had a key role in their day-to-day lives.

Alcatraz became part of the Golden Gate National Recreational Area in 1972, and visitors to Alcatraz today usually find the old cell blocks and the stories of prison life most absorbing, but the traces of the Indian occupation of 1969–71 are still there. You can see where the warden's house once stood, accidentally destroyed by fire during the occupation in 1970. The large, four-story building perched on the rocks above the bay, up the roadway to the right of the chapel, was originally built as an army barracks and later became an apartment house for federal correctional officers and their families. Here, many of the Indian families stayed during the occupation. You can stand on Alcatraz, with the cold wind blowing through your hair and breathing the salt air, and imagine what it must have been like to be making a statement for your people that had waited for a century to be made.

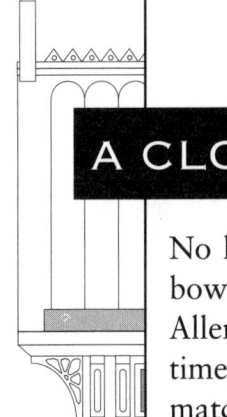

A CLOSE UP — LEGACY OF THE ALCATRAZ OCCUPATION

No longer willing to be seen as bronzed warriors with heads bowed in defeat, in the words of Paul Chaat Smith and Robert Allen Warrior, ". . . Indian people, for a brief and exhilarating time, staged a campaign of resistance and introspection unmatched in this century." This period of time spanned from November 1969 to May 1973, growing from the Red Power activism of the early 1960s. The campaign's stages included the student landing at Alcatraz, the Bureau of Indian Affairs occupation in Washington, and the siege at Wounded Knee, South Dakota in May 1973.

More recently, in 1990, 100 years had passed since the tragedy at Wounded Knee, and the Lakota people of South Dakota, as well as Native Americans throughout the country, took that opportunity to look back in time, to honor those who died there, and to observe its symbolism for all American Indians,

Alcatraz Proved-A-Point by Native American artist "Indian Joe" Morris evokes the sense of unity among Indians of All Tribes—some of whose names form the border. The symbols of another civilization surround them—the Golden Gate Bridge, the piers of San Francisco—as they salute the power of their past and their culture from the Rock. (From the Photo Collections of Golden Gate National Recreation Area National Park Service No. GoGA-2403.25, Indian Joe Morris Collection)

with whom they share a tremendous sense of loss, hurt, sorrow, and defeat. Many Lakota have also taken hope for a celebration of their unique culture, taking resolve to mend their sacred hoop, the Lakota Nation's traditional perception of identity, unity, and being in the integrated universe.

Today, through a reawakening of spiritual beliefs, values, and language, many believe that the sacred hoop of the Lakota and the cherished cultures of all American Indians—broken at the time of Wounded Knee—at last are being healed.

PRESERVING IT FOR THE FUTURE

In 1972, Alcatraz Island became part of the Golden Gate National Recreation Area (GGNRA) administered by the National Park Service. More than 750,000 visitors come to Alcatraz each year, and the Park Service is pursuing the goals of maintaining and restoring the historic buildings and providing further opportunities for public enjoyment. Future plans call for removal of much of the debris and other safety hazards, and opening additional areas of the island for public use and recreation.

EXPLORING ◆ FURTHER

Books about Alcatraz and the Indian Movement

Cheatham, Kae. *Dennis Banks: Native American Activist*. Native American Biographies. Springfield, N.J.: Enslow Publishing, 1997.

Eagle, Adam Fortunate. *Alcatraz! Alcatraz! The Indian Occupation, 1969–1971*. Photos by Ilka Hartmann. California Indian Series. Berkeley, Calif.: Heyday Books, 1992.

Johnson, Troy R. *The Occupation of Alcatraz: Indian Self-Determination and the Rise of Indian Activism*. Urbana, Ill.: University of Illinois Press, 1996.

Nielsen, Nancy. *Reformers and Activists*. New York: Facts On File, 1997.

Smith, Paul Chaat, and Robert Allen Warrior. *Like a Hurricane: The Indian Movement from Alcatraz to Wounded Knee*. New York: The New Press, 1996.

Related Place

Point Reyes National Seashore
State Highway 1
Point Reyes Station, CA 94956-9799
(415) 663-1092

Located near Olema on the California coast, this seashore area, with its ocean breakers, open grasslands, bushy hillsides, and forested ridges, was home to the Coast Miwok. Visitors can participate in seasonal celebrations of Coast Miwok culture and can tour an authentic replica of a Coast Miwok village.

MORE PLACES TO VISIT

The chapters of this book explore only a few of the dozens of historic sites that commemorate American Indians and their contributions and experiences. Following is a partial list of additional historical places that reflect the varied American Indian experience in the United States.

Mid-Atlantic States

Ganondagan State Historic Site
1488 Victor-Holcomb Road
Victor, NY 14565
(716) 924-5848

Partially restored Seneca town in the Genesee valley, where thousands of Seneca lived 300 years ago. Three trails marked with illustrated signs explore the significance of plant life to the Seneca, as well as Haudenosaunee (Iroquois) customs and beliefs. Interpretation covers the destruction of Ganondagan, or Town of Peace, by a large French army from Canada that attacked in 1687. The visitor center offers exhibits that explore the importance of the clan system to the Haudenosaunee and provides a 27-minute video on Ganondagan's history. Future plans include reconstruction of a Haudenosaunee longhouse.

Midwestern States

Mound City Group National Monument
16062 State Route 104
Chillicothe, OH 45601
(614) 774-1125

Mound City preserves 23 of the extraordinary earthwork mounds built by the Hopewell Indians of Ohio 2,000 years ago. These smooth, grassy, conical mounds served as burial sites for the Hopewell and remain preserved, enclosed by a 4-foot high earthwork wall. They range in height from 3 to 17 feet and serve as evidence of the sophisticated culture of the Native Americans who lived in the Midwest between 200 B.C. and A.D. 500. The mound closest to the visitor center has been excavated and glassed in, showing the layers of sand, gravel, topsoil, and river rock typical of the mound structure.

Northeastern States

Plimoth Plantation
Wampanoag Indian Program
P.O. Box 1620
Plymouth, MA 02360
(617) 746-1622

An outdoor living history museum that displays Native American artifacts from the colonial period and re-creates the life of a 1620s Wampanoag family that lived at Plymouth.

Northwestern States

Bering Land Bridge National Preserve
P.O. Box 220
Nome, AK 99762
(907) 443-2522

This preserve marks the spot where a land bridge connected the continents of Asia and North America more than 13,000 years ago. The land bridge itself is now covered by the waters of the Chukchi Sea and the Bering Sea. During the Ice Age, these waters receded enough to provide a land bridge traveled by people and animals and the plants they brought with them. Scientists have found evidence that prehistoric Asian hunters used this route to the New World, where many explored widely and settled. Inuit currently live in this area, pursuing subsistence lifestyles and managing reindeer herds. Usually accessed by air service out of Nome and Kotzebue, this highly isolated region requires extensive preparation and attention to safety and supplies for those who venture to visit.

Southeastern States

New Echota Historic Site
1211 Chatsworth Highway, N.E.
Calhoun, GA 30701
(404) 629-8151

New Echota is a reconstruction of the 1825 Cherokee capital town, a thriving settlement before 15,000 Cherokee were forced to move to Indian Territory in 1838—they traveled what became known as the Trail of Tears. Historical markers tell of the Trail of Tears, and the site features a restoration of the office of the *Cherokee Phoenix*, the Cherokee-language newspaper printed using type based on the alphabet developed by Sequoyah.

Southwestern States

San Antonio Missions National Historical Park
2202 Roosevelt Avenue
San Antonio, TX 78210
(210) 932-1001 or 534-8833

This group of four mission sites interprets the Spanish influence in Texas, including its effect on surrounding Native Americans from diverse hunting and gathering bands, described collectively by historians as Coahuiltecans [kwa weel TAY kens]. In return for receiving religious instruction at these missions, Indians were taught new techniques and skills in ranching, farming, and carpentry and were protected from members of other, more aggressive tribes. Visitors can follow a "Mission Trail," which is signed on the streets and for which a map is available from the Visitor Information Center. The missions include Mission Concepción (Misión Nuestra Señora de la Purísima Concepción, founded in 1731), Mission San José (Misión San José y San Miguel de Aguayo, 1720), Mission San Juan (Misión San Juan Capistrano, 1731), and Mission Espada (Misión San Francisco de la Espada, 1731).

Mission San Antonio de Valero ("the Alamo," founded in 1718) is located at the head of the Mission Trail, in downtown San Antonio.

Navajo National Monument
H.C. 71, Box 3
Tonalea, AZ 86044-9704
(520) 672-2366
Fax: (520) 672-2345

Betatakin, Keet Seel, and Inscription House (closed since 1968 due to its fragility) are three cliff dwellings of the Kayenta Anasazi. Proclaimed a national monument on March 20, 1909, the park's headquarters is located on 244.59 acres of tribal land adjacent to the Betatakin section (used by agreement of May 1962). A right-of-way of 4.59 acres was granted to the Park Service by the Navajo in 1977.

Western States

Big Hole National Battlefield
P.O. Box 237
Wisdom, MT 59761
(406) 689-3155

Located between Missoula and Butte in western Montana, this 655-acre tract of grassland and pine forest commemorates the bloody August 1877 battle between 800 members of the Nez Perce tribe and U.S. troops. Colonel John Gibbon, commanding the Seventh Infantry, attacked a sleeping encampment of Nez Perce on their way to join the Crow Indians in Montana. Led by Chief Joseph, an able and courageous leader, the Nez Perce had hoped to avoid confinement on a reservation in Idaho. Although victorious at the Big Hole battle, the Nez Perce strength was greatly depleted and less than two months later they surrendered to federal troops. A small museum at the battlefield exhibits Indian artifacts from the battle, including a coat owned by Chief Joseph. Visitors can follow foot trails to the battle site, the camp where the Indians slept, and the woods where the Native American warriors fought the U.S. troops for 24 hours.

Grand Teton National Park
Colter Bay Indian Art Museum
Moose, WY 83013
(307) 739-3300 (visitor's center)

This museum has a collection of 1,500 items related to Native American history and culture, with an emphasis on the tribes who inhabited the northern plains at the end of the 1800s and early 1900s. During the summer months films are shown and crafts demonstrations held.

MORE READING SOURCES

In addition to the suggested readings at the end of each chapter of this book, the following list provides additional suggestions for exploring further.

Andrews, Elaine K. *Indians of the Plains*. The First Americans Series. New York: Facts On File, 1992.

Begay, Shonto. *Navajo: Voices and Vision Across the Mesa*. New York: Scholastic, 1995.

Brown, Dee. *Bury My Heart at Wounded Knee—An Indian History of the American West*. New York: Holt, Rinehart, and Winston, 1970.

Calloway, Colin G. *Indians of the Northeast*. The First Americans Series. New York: Facts On File, 1991.

Duvall, Jill. *The Mohawk*. New True Books. Chicago: Children's Press, 1991.

———. *Seneca*. New True Books, Chicago: Children's Press, 1991.

Eagle/Walking Turtle. *Indian America: A Traveler's Companion*. 2d ed. Santa Fe: John Muir Press, 1991.

Force, Roland W., and Maryanne Tefft Force. *The American Indians*. New York: Chelsea House, 1991.

Gibson, Arrell Morgan. *The American Indian: Prehistory to the Present*. Lexington, Mass.: D.C. Heath and Co., 1980.

Hagman, Ruth. *The Crow*. New True Books. Chicago: Children's Press, 1990.

Hunt, Norman Bancroft. *North American Indians: The Life and Culture of the Native American*. Illustrated by Michael Codd. London: Brian Trodd Publishing House Limited, 1991.

Hoig, Stanley. *Night of the Cruel Moon: Cherokee Removal and the Trail of Tears*. New York: Facts On File, 1996.

Iverson, Peter J. *Navajos.* Indians of North America. New York: Chelsea House, 1990.

Josephy, Alvin M., Jr. *500 Nations: An Illustrated History of North American Indians.* New York: Alfred A. Knopf, 1994.

———. *The Indian Heritage of America.* New York: Alfred A. Knopf, 1968.

Josephy, Alvin M., Jr., Trudy Thomas, and Jeanne Eder. *Wounded Knee: Lest We Forget.* Cody, Wyo.: Buffalo Historical Center, 1990.

Keyworth, C. L. *California Indians.* The First Americans Series. New York: Facts On File, 1991.

Lawlor, Laurie. *Shadow Catcher: The Life and Works of Edward S. Curtis.* New York: Walker, 1994.

Liptak, Karen. *Indians of the Pacific Northwest.* The First Americans Series. New York: Facts On File, 1991.

———. *Indians of the Southwest.* The First Americans Series. New York: Facts On File, 1991.

———. *North American Indian Tribal Chiefs.* First Book. New York: Franklin Watts, 1992.

Lodge, Sally. *The Cheyenne.* Illustrated by Luciano Lazzarino. Native American People. Vero Beach, Fla.: Rourke Publications, 1990.

McCall, Barbara A. *The Apaches.* Illustrated by Luciano Lazzarino. Native American People. Vero Beach, Fla.: Rourke Publications, 1990.

Mankiller, Wilma. *A Chief and Her People: An Autobiography by the Principal Chief of the Cherokee Nation.* New York: St. Martin's Press, 1993.

Morgan, Ted. *Wilderness At Dawn: The Settling of the North American Continent.* New York: Simon & Schuster, 1993.

Neihardt, John G. *Black Elk Speaks.* New York: Simon & Schuster, 1959.

Olson, James Stuart. *The Ethnic Dimension in American History.* 2d ed. New York: St. Martin's Press, 1994.

Ourada, Patricia K. *The Menominee.* Edited by Frank W. Porter. New York: Chelsea House, 1990.

Quiri, Patricia Ryan. *The Algonquians.* New York: Franklin Watts, 1992.

Smith, Carter, ed. *Native Americans of the West: A Sourcebook on the American West.* American Albums from the Collections of the Library of Congress. Brookfield, Mass.: Millbrook Press, 1992.

Sherrow Victoria. *Indians of the Plateau and Great Basin.* The First Americans Series. New York: Facts On File, 1992.

Woodward, Grace Steele. *The Cherokees.* Norman: University of Oklahoma Press, 1963.

Younkin, Paula. *Indians of the Arctic and Subarctic.* The First Americans Series. New York: Facts On File, 1992.

INDEX

> **Boldface** page references indicate main headings. *Italic* page references indicate illustrations. The letter *m* following a page reference indicates a map or diagram.

A

Alabama 13, 32
Alabama River 33
Alaska xiii
Albuquerque, New Mexico 1, 3, 6, 11
Albuquerque Open Space Division 12
Alcatraz Island *110*
    occupation by Indian activists 107, **108–12**, *109, 111, 113*
    management of 114
Alcatraz Proved a Point 113
Aleut 73, 74, *75*, 78
American Indian Movement (AIM) 106, 111
American Indian Studies Center (at UCLA) 108
Anasazi, basket maker 40–41, *54*
Anasazi, Chaco Culture 51, **52–61**, *53, 55, 56, 57, 59, 61m* See also Aztec Ruins National Monument
    Classic period 55
    comparisons with Northern San Juan culture 57
    outlier communities 57–58, 63
    trade 57–58

Anasazi, Kayenta 119
Anasazi, Northern San Juan culture 14, **40–47**, *41, 42, 43, 44, 45, 46m, 47, 49m,* 50–51, 63 See also Aztec Ruins National Monument
Anasazi Heritage Center, Bureau of Land Management 50, 51
Arikara 64, 66–67
    and European explorers 67–68
Arizona 48
Aztec Ruins National Monument 62

B

Badlands National Park 105–6
Balcony House 44, *44*
Baranof Island 74, 77
Baranov, Alexander 72
Bartram, William 34
Battle of Sitka 72, 73, **74–75**, 77, 81
Bear Paw Mountains 102
Benteen, Frederick W. 87
Bering Land Bridge National Preserve 117–18
Bering Sea xiii, xiv, xv, 117
Bering Strait xiii, 27, 54

Big Foot (Lakota chief) 104
Big Hole National Battlefield 93, 119
Big Hole Valley 100–1, *101*
Big Thunder 98
Birch Creek, Idaho 102
Black Hills 86, 89, 92
Black Volcano 1, *5m*
Boca Negra Canyon *5m*, 11
Bodmer, Karl 68
Bond Volcano 1, *5m*
Brady, John 80
Bridgeport, Alabama 13
Brim (emperor of the Creek) 31
British traders and settlers
    and Creek Indians 30–32
Brown, Dee Alexander 73
Buffalo Bill Historical Center 71
Buffalo Bird Woman (Grace Henry) *70*
Bureau of Indian Affairs 111–12
Butte Volcano 1, *5m*

C

Cahokia Mounds Museum Society 23

Cahokia Mounds State Historic Site **15–23**, *17*, *18*, *19*, *22*
   history of site 15–19, 21
   map 20*m*
   preservation 23
Cahokians *See* Mississippian culture
California 6, 8 *See also* Alcatraz Island; Fort Ross State Historic Park; Point Reyes National Seashore
Camas Meadows, Idaho 102
Capitol Reef National Park 14
Casa Chaquita 52
Casa Rinconada 52, 58, **59–60**, *59*, 61*m*
Catholicism 9–10
Catlin, George 68
Chaco Canyon 51, 52, 53, 54, 57, 58, 59, 60, 61*m*, 63
Chaco Culture National Historical Park **52–61**, *53*, *55*, *56*, *57*, *59*
   establishment of 60
   map 61*m*
Chapin Mesa 49*m See also* Mesa Verde National Park
Chapin Mesa Museum 48, 49*m*
Charleston, South Carolina 31
Cherokee 32, 118
Cherokee Phoenix 118
Cherry Blossom Festival 36
Chetro Ketl 52, 55–56, *56*, *59*, 60, 61*m*
Cheyenne, Northern 83, 85, 88, 89
Cheyenne, Southern 85
Cheyenne River Indian Reservation 104
Chickasaw 31, 32
Chilkat robes 76, **79–80**, *79*
Chillicothe, Ohio 19, 23
Chippewa 85
Choctaw 32
Clark, William 64, 68, 97–98
Clearwater River 96, 100
cliff dwellings 39, 40, **42–47**, *58*
Cliff Palace *43*, *44*, 49*m*
Coahuiltecans 118
Collinsville, Illinois 15

Colorado 39 *See also* Dominguez and Escalante Ruins; Anasazi Heritage Center
Colorado River 54
Columbus, Christopher xiii
Colville Reservation 103
conquistadores 9
Cornfield Mound (Ocmulgee) 29, 33*m*
Crazy Horse 83, 85, 86, 89, 92
Crazy Horse Memorial 92
Creek Nation 26, **29–34** *See also* Mississippian culture
   battle of Horseshoe Bend 38
   conflicts with European settlers 31–33
   design of towns 29–30
   and fur trade 31
   at Ocmulgee 29, 31
   political system 29
   and Trail of Tears 33–34, 38
Creek War of 1813–14 38
Crook, George 86
Crow 85
Crow Agency, Montana 83
Currey, George 98
Custer Battlefield National Monument *See* Little Bighorn National Monument
Custer, George Armstrong 83, 85, 86

D

Dakota Territory 89
Davis, California 111
De Soto, Hernando xiv, 29
Defoe, Daniel xv
Dolores, Colorado 50
Dominguez and Escalante Ruins 50–51, 63
Dull Knife 89
Dupo, Illinois 18
Duran y Chaves, Fernando 9

E

earth lodges 27, 34–36, 67, **68–70**, *69*

Eastman, Charles A. 104
El Morro National Monument 62–63
Escalante Ruin *See* Dominguez and Escalante Ruins

F

Federal Bureau of Investigation (FBI) 110
Forsyth, James W. 104
Fort Buford, North Dakota 90
Fort Leavenworth 102
Fort Ross State Historic Park 81–82
Four Bears 65
Four Corners region 14, 54, 56–57 *See also* Chaco Culture National Historic Park; Mesa Verde National Park; Petroglyph National Monument
France
   involvement in Yamasee War 31
Fremont culture 14
Fremont River Canyon 14
Funeral Mound (Ocmulgee) 27–28

G

Gall (Lakota chief) 85
Ganondagan State Historic Site 116
Georgia 27 *See also* Ocmulgee Mounds National Monument
   establishment of colony 31
   transfer of land from Creek Indians to United States 32, 38
Ghost Dance movement 103–4
Gibbon, John 86, 100, 119
gold mining 86, 98
Golden Gate National Recreation Area 107, 112
Grand Plaza (Cahokia) 17, 20
Grand Teton National Park 120

Greasy Grass 86
Great Basin Abstract 6
Great Kiva (Casa Rinconada)
    52, 58, **59–60,** *59*
Great Kiva (Chetro Ketl) 55
Great Kiva (Lowry Ruins) 51
Great Pueblo 40, 44
Great Temple Mound 26,
    28–29, 33*m*, 36

H

Haida 72, 75
Haney, Woodrow, Sr. 26, 36
Harper's Ferry, Indiana 24
Haudenosaunee 116
Hell's Canyon 98
Hidatsa 64, 65, 66, 67, 69, 70
    earth lodges 68–70
    and European explorers
        67–68
    map of trading patterns
        66*m*
Hopewell Culture National
    Historical Park 19, **23–24**
Hopi 39, 40, 47
Horseshoe Bend National
    Military Park 38
Hovenweep 63
Howard, Oliver O. 99–100,
    102
Hudson's Bay Company 67
Hungo Pavi 52

I

Idaho *See* Nez Perce National
    Historical Park
Illinois 15, 16
Indian Movement *See* American Indian Movement
    (AIM); Red Power
Indian Petroglyph State Park
    11
Indian Territory *See* Oklahoma
Indians of All Tribes 108,
    109, *113*
Iroquois *See* Haudenosaunee
Iseminger, William R. 16

J

JA Volcano 1, 5*m*
Jackson, Andrew 38
Jackson, William Henry 44
Jamestown, Virginia 30
Jornada culture 13
Joseph (Nez Perce chief) 94,
    99, 102–3, 119
Josephy, Alvin M., Jr. 40

K

Kamiah, Idaho 96
Katlian 73, **74,** *75,* 78
Kayasha Pomo 82
Kidder, Alfred V. 52
Kin Kletso 52, 55, *55,* 56,
    61*m*
kivas 42, 45, 46*m*, 47,
    55–56, 58, **59–60,** 62
Knife River Indian Villages
    National Historic Site
    64–70, *65, 67, 69*
    establishment of 70
    map of trading patterns
        66*m*

L

Lakota 83, 85, 86
    and Crow 85
    conflicts with whites
        86–90
    Ghost Dancers and
        Wounded Knee
        Massacre 104, 106
"Lamar" culture 29, *30 See
    also* Creek Nation
Las Imagenes National Archaeological District 11
Last Stand Hill 85
Lebanon, Illinois 18
Lesser Temple Mound 26, **29,**
    33*m*
Lewis, Meriwether 64, 68,
    97–98
Like a Hurricane 108
Little Bighorn, Battle of the
    83, 84, 85, **86–87,** 88, 102
Little Bighorn Battlefield National Monument 84, 87
    establishment of 90
    map 91*m*
Little Bighorn River 86
Little Wolf 89
Long House 44
Looking Glass 100, 102
Lowry Ruins *See* Dominguez
    and Escalante Ruins

M

Mackenzie, Ranald S. 89
Macon Plateau xiv, 26, 31
    See also Ocmulgee National
        Monument
Mandan 64, 65, 67, 69, 70
    earth lodges 68–70
    and European explorers
        67–68
    map of trading patterns
        66*m*
    migration to Upper Missouri 66
"Manifest Destiny" 98
Mason, Charles 44
Mayans 21
McPhee Dam and Reservoir
    50
Meltzer, David J. xiii
Mesa Prieta 1, 5*m*
Mesa Verde National Park
    39–48, *41, 42, 43, 44, 45,
    47*
    establishment of 48
    map of Chapin Mesa
        49*m*
    map of Spruce Tree
        House 46*m*
    museum 48
    visitors' center 48
    wildlife 41–42
Mesoamericans 58
metates 4
Mexico, Gulf of 8
Miles, Nelson A. 102
missionaries 9
Mississippi Delta region 18
Mississippi River 15, 16, 17,
    31
Mississippian culture
    at Cahokia, **16–17,** *17,*
        18–22, *18, 19, 20m, 22*

at Ocmulgee 18, 26, **27–29, 34–36**
at Russell Cave 13
Missouri River 15, *65*, 66, 67
Mitchell, Illinois 18
Mogollon culture 13, 54
Monks Mound *17*, *18*, **19**, *19*, 20, 21
Montana *See* Big Hole National Battlefield; Little Bighorn Battlefield National Monument; Nez Perce National Historical Park
Montgomery, Alabama 33
Morris, "Indian Joe" 110, *112*
Mound City Group, 24, 116–17 *See also* Hopewell Culture National Historical Park
mounds, earthen
   at Cahokia State Historic Site xiv, xv, *17*, **18–19**, *18*, *20m*, 23
   at Effigy Mounds National Monument 24
   at Hopewell Culture National Historic Park 23–24
   at Ocmulgee National Monument 25, *26*, **27–29**, *30*, *33m*, **34–36**
Muskogean Indians *See also* Mississippian culture
   earth lodge at Ocmulgee 34–36

N

Nageezi, New Mexico v*m*, 52, 53
Natchez (tribe) 32
National Park Service 60, 93
Native American Art Gallery, Ocmulgee National Monument Museum 37
Native American University 111
Navajo 40
Navajo National Monument 119

Neva 74
Nevada 6
New Echota Historic Site 118
   See also Trail of Tears
New Mexico *See* Petroglyph National Monument; Chaco Culture National Historic Park; Mesa Verde National Park; Aztec Ruins National Monument; El Morro National Monument; Three Rivers Petroglyph Site
New Mexico, University of 11
Nez Perce 93, 94, *94*
   Coyote stories 95–96, *95*, *96*
   land cessions 98
   and horses 97
   non-treaty Nez Perce 99–103, *99m*, *101*, 119
   village life 97
Nez Perce National Historical Park 93–105
   establishment of 104
   map of Nez Perce National Historic Trail *99m*
Nimiipu *See* Nez Perce
Nixon, Richard 111
North Dakota *See* Knife River Indian Villages National Historic Site

O

Oakes, Richard 108, 110
Oakes, Yvonne 110
Ocmulgee National Monument v*m*, **25–37**, *26*, *28*, *30*, *33m*
   and Creek Nation 29–30, 31
   earliest inhabitants 27
   earthen mounds 27–29
   earth lodge 34–36
   visitors' facilities 37
Ocmulgee River 26, 27, 31, 35
Oglethorpe, James 31
Ohio River 66
Oklahoma
   and Creek Indians 33–34

Oregon *See* Nez Perce National Historical Park

P

Paiute 103
People of the Macon Plateau 37
Petroglyph National Monument **1–13** *See also* petroglyphs; Pueblo culture
   establishment of 11
   management 12
   map *5m*
   natural history of site 3–4
   visitors' centers 12
petroglyphs **6–11**, *7*, *8*, *62*, *63*
   dating 6
   imagery 3, 6–10
pictographs xiv, 11
Piedras Marcadas (pueblo) 8
Piedras Marcadas Canyon 1, *2*, *5m*
pit houses *41*
Pleistocene Era xiii–xiv
Plimoth Plantation 117
Point Reyes National Seashore 115
Powder River, Red Fork of the 89
Pueblo Bonito 52, *53*, 54, 59, 60
Pueblo culture 3, 4, 6–10
Pueblo del Arroyo 52, 56, *57*, 60, *61m*
Pueblo I phase (of petroglyphs) 6
pueblos (buildings) xiv, 8, 9, 54, *58*, *63* *See also names of specific pueblos*

R

Red Power
   occupation of Alcatraz Island **108–12**, *109*, *111*
   occupation of BIA headquarters 111–12
   siege at Wounded Knee 111, *112*

Red Stick Creek 38 *See also* Creek Nation
Redoubt St. Michael 74
Reno, Marcus A. 87
Reno's Retreat 88
Revolutionary War 32
Rinconada Canyon 4, *5m*, 8
Rinconada Mesa 4
Rio Grande 3, 4, *5m*, 8, 9, 27
Rio Puerco 4
Rousseau, Jean-Jacques xv
Russell Cave National Monument 13
Russian traders 72, 74–75, 78, 82
Russian-American Company 72, 74

S

Sacagawea 64
Salmon River 98, 100
San Antonio Missions National Historical Park 118–19
San Francisco Bay 107, 108
San Juan Basin 58
Savannah River 32
Scioto River 24
Seneca 116
Sequoyah 118
Shawnee 32
Shee Atika 74
Shoshone 102
Sitka National Historical Park **72–81**, *73*, *75*, *77*, *79*
  establishment of 81
  totem poles preserved in 78, 80–81
Sitka Sound 78
Sitting Bull 83, 85, 86, 89–90, 102
Smith, Paul Chaat 108, 112
Snake River 98, 100
South Carolina 31
South Dakota 66 *See also* Crazy Horse Memorial; Dakota Territory
Spain
  and Pueblo culture 9–10

involvement in Yamasee War 31
Spruce Tree House 44, **45–47**, *45*, *46m*, *47*, *49m*
St. Louis, Missouri 15, 16, 18
Stonehenge 21, 22
Sun Dance 86

T

Tallapoosa River 38
Taos Pueblo 111
Tecumseh 32
Terry, Alfred H. 86
Tewa 39
Three Affiliated Tribes *See* Arikara; Hidatsa; Mandan
Three Rivers Petroglyph Site 13
tipis 97
Tlingit Kikadsi clan 72, 73, 74, *75 See also* Chilkat robes
  battle of Sitka 72, 73, **74–75**, 77
  fishing 75–76
  trade 76, 80
  potlatch ceremony 76
  totems and carvings 76–78
Torrey, Utah 14
totem poles 72, 73, 76, **77–78**
  destruction by missionaries 78
  at Sitka National Historical Park 80–81
Trail of Tears 33–34
Tsimshian 75
Tuekakas 99
Twin Mounds 17, 19, 20
Two Moons 87, 111

U

U.S. Infantry, 39th 38
Una Vida 52
United Nations Educational, Scientific and Cultural Organization (UNESCO) 48

Utah 14, 58 *See* Capitol Reef National Park
Utley, Robert M. 84

V

Verendrye (explorers) 67
Vespucci, Amerigo xiii
Vikings xiii
Volcano City Park 11
Vulcan Volcano 1, *5m*

W

Wallowa Lake 98
Wampanoag 117
War of 1812 32, 38
Warrior, Robert Allen 108, 112
Washington *See* Nez Perce National Historical Park
West Mesa 3–4, 8, 9, 27
Wetherill, Richard 44
White Bird 88
White Bird Canyon 100
Woodhenge *17*, 20, **21–22**, *22*
Woodland culture xiv, 13
  Eastern 24
  Late 17
Wounded Knee Creek
  AIM demonstration at 111
  massacre of Lakota at 104, 106
Wounded Knee Monument 106
Wovoka 103
Wyoming *See* Buffalo Bill Historical Center

Y

Yamasee War 31 *See also* Creek Nation
Yellowstone River 102

Z

Ziolkowski, Korczak 92
Zuni 39, 40

THE AMERICAN INDIAN EXPERIENCE